Probability-Based Multi-objective Optimization for Material Selection

Maosheng Zheng · Jie Yu · Haipeng Teng ·
Ying Cui · Yi Wang

Probability-Based Multi-objective Optimization for Material Selection

Second Edition

Springer

Maosheng Zheng
School of Chemical Engineering
Northwest University
Xi'an, Shaanxi, China

Haipeng Teng
School of Chemical Engineering
Northwest University
Xi'an, Shaanxi, China

Yi Wang
School of Chemical Engineering
Northwest University
Xi'an, Shaanxi, China

Jie Yu
School of Life Science
Northwest University
Xi'an, Shaanxi, China

Ying Cui
China Railway Baoji Track Electrical
Equipment Inspection Co., Ltd.
Baoji, Shaanxi, China

ISBN 978-981-99-3938-1 ISBN 978-981-99-3939-8 (eBook)
https://doi.org/10.1007/978-981-99-3939-8

This Springer imprint is published by the registered company Springer Nature Singapore Pte Ltd.
The registered company address is: 152 Beach Road, #21-01/04 Gateway East, Singapore 189721, Singapore

Preface to the Second Edition

Since the publication of the first edition of *Probability-Based Multi-objective Optimization for Material Selection*, some subsequent works have been done, which promote us to have the possibility to publish the second edition of the book with new complements. Especially, the viewpoint of system theory in considering "simultaneous optimization of multiple objectives" is put forward.

In addition to the original materials of the first edition, the fuzzy-based probabilistic multi-objective optimization, cluster analysis of multiple objectives, treatments of portfolio investment, and multi-objective shortest path problem by means of probability-based multi-objective optimization are all developed in the second edition. Adjustment of arrangement of some chapters is involved. Besides, some minor supplements and error corrections of the first edition are conducted.

Nevertheless, the aim of this book is still to cast a brick to attract jade and would make its contributions to relevant fields as a paving stone.

The main purpose of this book is to provide a rational way for material selection in viewpoint of system theory and in the spirit of probability theory with reasonable physical essence. It is our great pleasure if the readers including scientists, engineers, postgraduate, and advanced undergraduate in the relevant fields could gain valuable information from this book.

The contents of the second edition are as following:

Chapter 1 describes the history and current status of material selection with multi-objective optimization briefly;

Chapter 2 reviews and summarizes the previous methods for material selection with multi-objective optimization mainly, including Farag comprehensive method, analytic hierarchy process (AHP), Vlšekriterijumsko KOmpromisno Rangiranje (VIKOR), technique of ranking preferences by similarity to the ideal solution (TOPSIS), multi-objective optimization (MOO) on the basis of ratio analysis (MOORA), Ashby's method, etc.;

Chapter 3 illuminates the fundamental principle and concepts of probability-based multi-objective optimization for material selection in viewpoint of system theory and in the spirit of probability theory;

Chapter 4 presents the robustness evaluation with probability-based multi-objective optimization in condition of the utility of response with uncertainty;

Chapter 5 describes the extension of probability-based multi-objective optimization in condition of the utility with desirable value;

Chapter 6 explains the hybrids of probability-based multi-objective optimization with experiment design methodologies, i.e., orthogonal experimental design, response surface design, and uniform experimental design;

Chapter 7 illuminates discretization of simplified evaluation in probability-based multi-objective optimization by means of GLP and uniform experimental design;

Chapter 8 presents the fuzzy-based probabilistic multi-objective optimization;

Chapter 9 describes the cluster analysis of multiple objectives in probability-based multi-objective optimization;

Chapter 10 states the applications of probability-based multi-objective optimization beyond material selection;

Chapter 11 shows the treatment of portfolio investment by means of probability-based multi-objective optimization;

Chapter 12 displays the treatment of multi-objective shortest path problem by means of probability-based multi-objective optimization;

Chapter 13 discusses on preferable probability, discretization, error analysis, hybrid of sequential uniform design with PMOO, and weighting factor evaluation;

Chapter 14 gives a comprehensive summary of the probability-based multi-objective optimization.

Professor Kaiping Liu is acknowledged for his valuable discussion in improving the texts. The authors hope that this book can inspire people's enthusiasm for new exploration on multi-objective optimization problems and play the role of paving the way.

Xi'an, China

Maosheng Zheng
Jie Yu
Haipeng Teng
Ying Cui
Yi Wang

Preface to the First Edition

Material is a magical field. The author group has been devoted to material researches since early 1990s, and especially the first author has lectured the relevant courses for about 30 years, a lot of scenes are experienced, which motivates the authors to pay more attention to realize the nature of material behavior, and to develop appropriate methodology and idea to characterize material performances including material selection due to the vast amount of materials and their performances, as well as wide application fields.

In 2019, a book entitled *Elastoplastic Behavior of Highly Ductile Materials* (authored by Zheng M., Yin Z., Teng H., Liu J. and Wang Y, Springer Press, Singapore) was published, which aimed to preach the benefits of high ductility of materials to safety of components in different respects of elastoplastic deformation and anti-failure, while the current book named *Probability-Based Multi-objective Optimization for Material Selection* focuses on appropriate choice and overall/comprehensive evaluation of material utility quantitatively for practical engineering application in viewpoint of probability theory systematically, which can be seen as a growing branch of applications and prospect assessments of materials with novel idea and methodology.

In order to understand the contents of this book thoroughly, the fundamental knowledge of probability theory and optimum theory is needed for researchers working in material fields to warm up.

Materials inevitably influenced daily lives of human kind from ancient time till now. Of course, in ancient time, the number of materials was very small, the need for material selection was quite rare, but today number of materials and their performances are quite vast. The chances for innovation that materials offer are equally immense now. It is only possible to select a proper material if there is a procedure for material selection rationally from the large number of material bank; simultaneously the material selection is usually related to the manufacturing process, cost, and environmental friendship in the entire lifetime, etc., therefore material selection is in fact not an easy task. This book illuminates the newly developed probability-based approach of multi-objective optimization for selecting materials systematically, which aims to deal with the relevant problems in material selection rationally.

In the treatment, a brand new concept of preferable probability is introduced to reflect the preferable degree of the utility of each material performance indicator, and the overall/total preferable probability of the candidate material is its unique and decisive index in the selection quantitatively. The combinations of this novel approach with the experimental design methodologies, including orthogonal experimental design, response surface methodology, and uniform experiment design, are all included; the robustness assessment of experimental results with dispersion is contained as well; the discretization treatment of complicated integral in the evaluation is presented. Ten chapters are devoted to describe the whole issues in details.

The main purpose of this book is to provide a rational way for material selection in viewpoint of probability theory with reasonable physical essence. It is our great pleasure if the readers including scientists, engineers, postgraduate, and advanced undergraduate in the relevant fields could gain valuable information from this book.

Chapter 1 describes the history and current status of material selection with multi-objective optimization briefly;

Chapter 2 reviews and summarizes the previous methods for material selection with multi-objective optimization mainly, including Farag comprehensive method, analytic hierarchy process (AHP), VIšekriterijumsko KOmpromisno Rangiranje (VIKOR), technique of ranking preferences by similarity to the ideal solution (TOPSIS), multi-objective optimization (MOO) on the basis of ratio analysis (MOORA), Ashby's method, etc.;

Chapter 3 illuminates the fundamental principle and concepts of probability-based multi-objective optimization for material selection;

Chapter 4 presents the extension of probability-based multi-objective optimization in condition of the utility with interval number and robust design of experiment;

Chapter 5 describes the extension of probability-based multi-objective optimization in condition of the utility with desirable value;

Chapter 6 explains the combinations of probability-based multi-objective optimization with experiment design methodologies, i.e., orthogonal experimental design, response surface design, and uniform experimental design;

Chapter 7 illuminates discretization treatment of complicated integral in assessing probability-based multi-objective optimization by means of good lattice point and uniform experimental design;

Chapter 8 states the applications of probability-based multi-objective optimization beyond material selection;

Chapter 9 gives a comprehensive summary of the probability-based multi-objective optimization;

Chapter 10 stresses some words to initiate further discussion for the related problems concerning preferable probability and its quantitative evaluation.

Thanks are given to Profs. Jianlong Zheng and Mingxin Tong and Mr. Xiaokang Shen for their continuing support in convening seminars and other respects; Mr. Zhijie Yang is acknowledged for his effort in conducting the calculation and analysis of multi-objective optimization of numerical control machining parameters for high efficiency and low carbon of example 6 in Sect. 7.5.

The authors wish this work will cast a brick to attract jade and would make its contributions to relevant fields as a paving stone.

Xi'an, China

Maosheng Zheng
Haipeng Teng
Jie Yu
Ying Cui
Yi Wang

Contents

About the Authors

Dr. Maosheng Zheng received his Bachelor's degree in Theoretical Physics from Northwest University, Xi'an, China, in 1983, Master's degree in Electronic Physics from Xidian University, Xi'an, China, in 1985, and Ph.D. degree in Materials Science and Engineering from Northwestern Polytechnic University, Xi'an, China, in 1992. In June 1994 to June 2006, He has been in School of Materials Science and Engineering, Xi'an Jiaotong University as Professor, and since June 2006, Dr. Zheng has been in School of Chemical Engineering, Northwest University, as Professor.

He focuses on the research of Materials Science and Technology. His recent research interests include material selection with multi-objective optimization, material technology, energy resource material and technology, renewable energy conversion and utilization, etc. Till 2023, he has published three monographs (first authored 2, Springer; co-authored, Science Press) more than 300 peer-reviewed papers in international journals and conferences, chaired "863" project and many projects of Science and Technology Ministry and Shaanxi Province of China. He is Member of China Energy Society, Vice Chairman of Shaanxi Corrosion and Protection Society, and served as Reviewer of several international journals. He was Recipient of State Council special allowance, second prize of Science and Technology Progress Award of China Aviation Industry Corporation, third prize of Science and Technology Progress Award of the Ministry of Education, second prize of Huo Yingdong Young Teacher (Research), The 2nd Youth Science and Technology Award of Shaanxi Province, second prize of the Natural Science Award of China Higher Education, respectively.

Jie Yu received her Bachelor's degree in polymer science and technology from Northwestern Polytechnic University, Xi'an, China, in 1987. In July 1987 to December 1992, Yu worked at Xi'an Far East Machinery Manufacturing Company as Assistant Engineer responsible for the design and processing of rubber and plastic parts; in January 1993 to September 2007, she served as Engineer in Department of Polymer Materials of Xi'an Jiaotong University; in September 2007 to August 2020, she worked as Senior Engineer, College of Life Sciences of Northwest University; since

August 2020, she served as Professor-leveled Senior Engineer at College of Life Sciences of Northwest University.

She focuses on design of experiment and optimization, research of polymer materials, and their applications in medicine. Till 2023, Yu has published one monograph (co-authored, Springer) and more than 60 scientific research papers, chaired and completed two key projects of Shaanxi Province, participated in the completion of five national and provincial key projects and two enterprise cooperation projects, obtained national invention patents three items, and two items have been transformed to serve the society, won one advanced worker of Northwest University, two science and technology awards of Shaanxi Provincial Higher Education Institutions, and one gold medal of Shaanxi Science and Technology Workers Innovation and Entrepreneurship Competition.

Dr. Haipeng Teng received his Bachelor's degree in Thermal and Power Engineering from Xi'an Jiaotong University, Xi'an, China, in 2002 and Ph.D. degree in Thermal and Power Engineering from Institute of Engineering Thermophysics of Chinese Academy of Sciences, Beijing, China, in 2011. In July 2002 to July 2006, Dr. Teng has been Teacher in College of Mechanical and Electrical Engineering, Shaanxi University of Science and Technology. In July 2011, he started as Lecturer in School of Chemical Engineering, Northwest University, and Associate Professor in 2014.

He focuses on the research of energy resource material, thermal energy conversion and utilization, renewable energy, etc. Till 2023, he has published 2 monograph (co-authored, Springer) and more than 40 peer-reviewed papers in international journals and conferences and completed more than ten scientific research projects.

Ying Cui received her Bachelor of Mechanical Engineering from Shaanxi University of Technology in July 1996, Master of Industrial Engineering from Sichuan University in 2014. In August 1996 to May 2018, she worked at Baoji Equipment Factory of China Railway Electrification Bureau Group Company, and since February 2007, she served as Senior Engineer. Since May 2018, Cui served as Chief Engineer and Senior Engineer of China Railway Baoji Railway Electrical Equipment Inspection Co. She is Engineering and Technical Expert of China Railway Electrification Bureau Group Co., Ltd., First-level Procurement Expert of China Railway Corporation, and a top talent in Baoji High-tech Zone. She has successively engaged in the design, manufacturing, and testing of electrified railways and urban rail contact net products, R&D and testing of mechanical components for electrified railways and urban rail transit; presided over the "Shaanxi Railway Electrical Equipment Testing Service Platform", "Rigid Suspension Anti-seize Device", "Straddle-type Monorail Contact Rail Power Supply System Product Research", more than 20 key scientific research projects, etc.; presided over the revision of the railway industry standard "Electrified Railway Busbars and Components", participated in the "Rail Transit Rigid Overhead Catenary Project Construction Quality Acceptance Standard' and other standards, published one monograph (co-authored, Springer) and more than ten papers, and obtained more than 40 patents. She won the first prize of Science and Technology Progress of China

Railway Engineering Corporation, a number of science and technology awards from the city and group companies.

Yi Wang received her Bachelor's degree in Mechanical Design, Manufacturing and Automation from Northwest A&F University, China, in 2014 and Master's degree in Mechanical Design and Theory from Xi'an Shiyou University, China, in 2017. Since June 2018, she has been in School of Chemical Engineering, Northwest University, as Assistant Engineer engaged mainly in Process Equipment and Control Engineering, and since October 2020, she served as Engineer. Till 2023, Wang has published two monographs (co-authored, Springer) and 15 peer-reviewed papers in international journals and conferences.

Chapter 1
History and Current Status of Material Selection with Multi-objective Optimization

Abstract The history and current status of material selection with multi-objective optimization is stated briefly; it aims to display the complexity in material selection in nowadays due to the large amount of materials and their performances, as well as the complex factors to be considered.

1.1 Brief Introduction

There are vastly 160,000 or more materials available to designers and engineers [1]. The persisting appearance of novel materials with new and exploitable properties extends the options continuously, which raises a problem that how designers and engineers choose their materials from the large material bank, which material is the best to suit their purpose? And how do they know?

The selection of material cannot be performed isolatedly without considering the correlated processing technology, by which the material is shaped, joined, and finished, and the relevant cost, as well as the effects of manufacturing and application on the surrounding environments. In nowadays, in almost everything from household products to automobiles, aircraft, or even space ship, from their form, texture, feel, color, beauty, and satisfaction of the products, many attributes are needed to be considered. Some aspects are even conflicting each other in design.

It was said that design problems are almost always open ended, which do not have a unique or "correct" solution, though some solutions look better than others distinctly. Different idea of design may lead to different consequence.

In practice, material selection has a long history, from the house building in ancient time to daily life products shopping of nowadays; it involves many objectives which are even conflicting.

This book illuminates the probability-based approach of multi-objective optimization for material selection systematically, which aims to provide a novel methodology through the forest of complex choices. The utility of material performance and process attributes are characterized by a new idea of preferable probability, which links up the whole material selection processes. The hybrids of probability-based approach with experimental designs and robustness of product processing are

M. Zheng et al., *Probability-Based Multi-objective Optimization for Material Selection*, https://doi.org/10.1007/978-981-99-3939-8_1

involved inevitably as well. The history and current status of material selection with multi-objective optimization is stated briefly, which contains dramatic and attractive stories in some sense.

1.2 Evolution of Material Selections

Throughout history, the ages of mankind are named by materials' type that was used in that time: stone, bronze, iron, and now "information time". And the famous personages were even buried with their materials; e.g., the First Emperor of Qin Dynasty (259–210BC) was buried with his Bronze Chariots and Horses and Terracotta Warriors, Agamemnon was buried with his Bronze Sword and Mask of Gold, and Viking chieftains was buried in their burial ships—each treasure showing the "high technology" of their days. If today a modern rich merchant is died, his titanium watch, or carbon nanotube-reinforced iPod, carbon fiber-reinforced fishing rod, their metal–matrix composite mountain bike, will accompany him, or all together, perhaps. Why so? This is the time with full of a large amount of materials.

This is not only an era with a large number of materials, but with varying properties. So, the material selection for specific usage for the designers and engineers is of significance including innovative design and improvement of their products.

Before 10,000 BC (the Stone Age), the materials used in prehistory time included ceramics and glasses, natural polymers, and composites. The weapons were always made of wood and flint, which were the marks of the peak of technology. Buildings and bridges were very natural, which are made of stone and wood. Till 4000–1000 BC (the Bronze Age) and 1000 BC to 1620 AD (the Iron Age), the explorations of rudimentary thermochemistry allowed the extraction of copper and bronze first and then iron, so technology and production got significant advances at that time.

Iron cast ironing technology (1620s) established the dominance of metals in engineering; since then, the rapid development of steels (1850 onward), light alloys (1940s), and special alloys consolidated their solid position. By the years of 1950s, the concept of "engineering materials" formed. Thereafter, the development of nonmetal materials occurred including cements, refractories, and glasses; and development of rubber, bakelite, and polyethylene among polymers was also very rapid though with small share in the total materials market. Since 1950, the polymer, high-performance ceramics and composite industries have grown rapidly. In the present day, the things accelerate with expanded need dramatically. The fascinating question is more serious in material selection from the respects of material performance, manufacture, environments, etc.

1.3 Evolution of Multi-objective Optimization

The process of optimizing a collection of objective functions systematically and simultaneously is called multi-objective optimization (MOO).

The historical origins of MOO can be traced back to 1103, it was the time of Song Dynasty in China, an architectural specialist Li Jie (J. Li) authored a book entitled "Yingzao Fashi" (Rules of Building) which systematically described the rule of ratio of height to width of wooden beam with 3:2 [2], which is an optimal ratio in comprehensive (overall) consideration of the strength and rigidity in viewpoint of modern material mechanics; it accounted out that the rate of 102 beams of 31 buildings of Song Dynasty with ratio of height to width over $\sqrt{2}$:1 was about 77.5% [2].

Other traces were correspondence between Nicolas Bernoulli (1687–1759) and Pierre Rémond de Montmort (1678–1719), discussing the St. Petersburg paradox [3, 4]. Until 1738, Daniel Bernoulli published his influential research on utility theory, the answer to the St. Petersburg paradox was available. The conclusion is that humans could make decisions with utility value instead of expected value of things. The implication of the utility value is that the highest utility value is the optimal chosen from the alternative in MOO problems [3].

In 1879, Pareto developed Pareto optimality [4], which was seen as main basis of modern MOO. Marler et al. pointed that a solution to a multi-objective problem is more like a concept than a definition in respect to a single-objective optimization [4]. Typically, a single global solution is not existed, and it is often necessary to decide a set of points which fulfill the predetermined definition for an optimum. An alternative to the idea of Pareto optimality and efficiency is the idea of a compromise solution. It entails to seek minimum of the difference between the potential optimal point and a utopia point. So the next thing is to search a solution which is a compromised solution and as close as possible to the utopia point. One difficulty with the idea of a compromised solution is the definition of the word "close". Besides, if different units are involved in the different objective functions, the Euclidean norm or a norm of any degree is not sufficient to reflect the closeness mathematically. As a result, the objective functions should be transformed such that they are dimensionless. The most commonly used method is scalarization for multi-objective optimization; however, the scaled factor in the scalarized function is still a puzzled issue. Additionally, the exponent p and the weights in Pareto optimality were continuously worried. Misinterpretation of the theoretical and practical meaning of the weights can lead to the process and result strange [4].

In 1947, von Neumann and Morgenstern published a book entitled "*Theory of Games and Economic Behavior*" to concrete a mathematical theory of economic and social organization, which forms the embryo of MOO [3]. Then in 1951, Kuhn and Tucker published the vector optimization concept for multiple objectives and Yu in 1973 proposed the compromised solution method to cope with MOO problems, a lot of work has been conducted on various applications, e.g., planning and transportation investment, development and planning econometrics, water resource management, environmental issues and public policy, etc.

In considering the subjective uncertainty, fuzzy numbers were incorporated into MOO for dealing with more extensive problems in 1970s after Bellman and Zadeh's fuzzy set.

While in 1951, Box and Wilson of Imperial Chemical Industries created the response surface methodology (RSM) and robust design [5, 6]. The application of RSM to chemical processes was described, which initiated the industrial applications of experimental design and attention of research in the field.

In Japan, Taguchi proposed "Taguchi Methods" with orthogonal arrays for experimental design to conduct the practical optimal problem and robust design as well, which gets successive achievements [7].

In the same period in China, Hua L. K. (Hua Loo Keng) led a group of assistants to conduct popularization of optimum seeking method and overall planning method (CPM + PERT) in industries through 26 provinces of China in 1960–70s [8–10], which pushed forward the development of technology and production rapidly and laid a solid foundation of optimization in China. The term CPM means "critical path method", and the meaning for "PERT" is Program Evaluation and Review Technique in the west in 1950s. CPM and PERT are two methods for planning, which appeared almost simultaneously in the late 1950s. With the rapid development of science and technology and production, many huge and complex scientific research and engineering projects have appeared. They have many processes and extensive cooperation and often require a lot of human, material, and financial resources. Therefore, how to organize them rationally and effectively, and make them coordinate with each other, under the limited resources, with the shortest time and the lowest cost, the best way to complete the whole project has become a prominent and important issue. Hua's optimum seeking method contained golden section method and Fibonacci search method mainly, which were popularized to folks by Hua's popularization at that time. Thereafter, the optimum seeking method was popularly used in industry due to promotion of Hua's group. During their promotion, Hua's books "Plain Talk on Optimum Seeking Method" and "Plain Talk on Overall Planning Method" were published, which not only popularized the basic knowledge and conducting folks' training, but also promoted the actual development of technique and production at that time dramatically. Of course, Hua's popularization broke a new way from theoretic research to practice. In this period, the orthogonal design was introduced and used in China as well [11, 12].

During Hua's popularization time, he specially stressed that the optimum seeking methods for two variables could be easily generalized to the situation of more variables. The better approach is to grasp the principal factors such as one or two variables in optimum seeking method in industrial production process so as to obtain a better production technology [10].

With the development of science and technology, these two methods (optimum seeking method and overall planning method) were not able to solve many practical problems. The golden ratio optimization method is an optimal method to deal with a single variable problem, which is almost very little case in practice, while orthogonal design is on basis of Latin square theory and group theory and can be used to conduct multifactor experiments. As a result, the number of trials is significantly reduced for all combinations of different levels of factors. However, the number of trials of orthogonal design is still too high and could not be facilitated for some expensive scientific or industrialized experiments. Later on, in 1978, Fang and Wang faced experimental design problem of a five-variable experiment with 18 levels for each variable and the total number of trials being limited not larger than 50 for missiles at that time. It is not possible to use orthogonal design to complete this design [11–13]. After hard work for a few months, Professors Fang and Wang put forward a new type of experimental designs known as "uniform experimental design (UED)". The new method was successfully applied to the experimental design of missiles. Thereafter, UED has been widely applied in China with a series of gratifying achievements. The UED belongs to the quasi-Monte Carlo methods or number-theoretical methods [11–14]. The quasi-Monte Carlo method, or number-theoretical method was successfully applied in approximate numerical calculations for multiple integrals. Recently, the probability-based approach for multi-objective optimization (PMOO) was developed to describe the "simultaneous optimization" of multiple objectives as an overall consideration in viewpoint of system theory, and it was further combined with uniform experimental design, orthogonal design, and response surface design to extend its application involving experimental design problems, as well as robustness evaluations and good results were obtained [15–19]. Other approaches, such as AHP, VIKOR, TOPSIS, MOORA, and Farag's comprehensive method, contain features of subjective factors and "additive" algorithm, which reflects the "simultaneous optimization" of multiple objectives insufficiently in viewpoint of system theory [15–19]. Ashby's method and Farag's comprehensive method consider more performance characteristics of material itself, which could be used for initial screen [1, 20].

1.4 Summary and Conclusions

The rapid development of material and the ways of products' usage force the selection of materials a multi-objective problem with conflicting requirements of performance and utility, including performance of material itself, manufacture process and its robustness, cost and environment friendship in its entire lifetime, etc.

As to all above attributes and objectives, proper approach of quantitative assessment for material selection is in need. The appropriate approach for such assessment must reflect the impersonal characteristics of performance and utility of material perfectly and comprehensively.

Theory roots in practice and serves practice; we hope this work will cast a brick to attract jade and would make its contributions to relevant fields as a paving stone.

References

1. M.F. Ashby, *Materials Selection in Mechanical Design*, 4th edn. (Butterworth–Heinemann, Burlington, MA, 2011)
2. L. Lao, *History of Material Mechanics in Ancient China* (National Defense University Press, Changsha, 1991), pp.110–119
3. G.H. Tzeng, J.J. Huang, *Multiple Attribute Decision Making Methods and applications* (CRC Press, Taylor & Francis Group, 2011)
4. R.T. Marler, J.S. Arora, Survey of multi-objective optimization methods for engineering. Struct. Multidisc. Optim. **26**, 369–395 (2004). https://doi.org/10.1007/s00158-003-0368-6
5. G.E.P. Box, K.B. Wilson, On the experimental attainment of optimum conditions. J. Roy. Stat. Soc. Ser. B (Methodol.) **13**(1), 1–45 (1951)
6. R.H. Myers, D.C. Montgomery, C.M. Anderson-Cook, *Response Surface Methodology, Process and Product Optimization Using Designed Experiments*, 4th edn. (Wiley, New Jersey, 2016)
7. T. Mori, *Taguchi Methods, Benefits, Impacts, Mathematics, Statistics, and Applications* (ASME Press, New York, 2011)
8. S. Gong, The life and work of famous Chinese mathematician Loo-keng Hua. Adv. Appl. Clifford Algebras **11**(S2), 9–20 (2001)
9. J. Hudeček, Hua Loo–Keng's popularization of mathematics and the cultural revolution. Endeavour **41**(3), 85–93 (2017)
10. L.K. Hua, Y. Wang, J.G.C. Heijmans, *Popularizing Mathematical Methods in the People's Republic of China: Some Personal Experiences* (Birkhäuser, Boston, 1989)
11. J. Fan, J. Pan, *Contemporary Experimental Design, Multivariate Analysis and Data Mining, Festschrift in Honour of Professor K. T. Fang* (Springer, Cham, 2020)
12. K.T. Fang, M.Q. Liu, H. Qin, Y.D. Zhou, *Theory and Application of Uniform Experimental Designs* (Science Press and Springer, Beijing, China, and Singapore, 2018)
13. Y. Wang, K.T. Fang, A note on uniform distribution and experimental design. Chin. Sci. Bull. **26**, 485–489 (1981)
14. Y. Wang, K.T. Fang, On number—theoretic method in statistics simulation. Sci. Chin. A **53**(1), 179–186 (2010)
15. M. Zheng, Y. Wang, H. Teng, An novel method based on probability theory for simultaneous optimization of multi-object orthogonal test design in material engineering. Kovove Mat. **60**(1), 45–53 (2022). https://doi.org/10.31577/km.2022.1.45
16. M. Zheng, Y. Wang, H. Teng, A new "intersection" method for multi-objective optimization in material selection. Tehnički Glas. **15**(4), 562–568 (2021). https://doi.org/10.31803/tg-20210901142449
17. M. Zheng, Y. Wang, H. Teng, Hybrid of the "intersection" algorithm for multi-objective optimization with response surface methodology and its application. Tehnički Glas. **16**(4), 454–457 (2022). https://doi.org/10.31803/tg-20210930051227
18. M. Zheng, H. Teng, Y. Wang, Robust design in material machining process on basis of probability multi-objective optimization. Materialwiss und Werkstofftech **54**(2), 180–185 (2023). https://doi.org/10.1002/mawe.202200162
19. M. Zheng, Y. Wang, H. Teng, A novel approach based on probability theory for material selection. Materialwiss und Werkstofftech **53**(6), 666–674 (2022). https://doi.org/10.1002/mawe.202100226
20. M.M. Farag, *Materials and Process Selection for Engineering Design*, 4th edn. (CRC Press, Taylor & Francis Group, New York, 2010), pp. 328–334

Chapter 2 Introduction to Multi-objective Optimization in Material Selections

Abstract It mainly summarizes and reviews the previous methods for material selection with multi-objective optimization, including Farag comprehensive method, analytic hierarchy process (AHP), Vlšekriterijumska Optimizacija I KOmpromisno Resenje (VIKOR), technique of ranking preferences by similarity to the ideal solution (TOPSIS), multi-objective optimization (MOO) on the basis of ratio analysis (MOORA), Ashby's method, etc.

2.1 Introduction

In practical life and engineering fields, everyone is a decision-maker, who is responsible to make choice for balance judgments among many objectives or attributes of things with his or her preferences for possible consequences or outcomes. So, decision making or optimization for multiple objectives or attributes is nothing new in some sense. However, it is not easy to make proper choice under condition of multiple objectives with conflicting tendency, even though we all have a lot of "practice" and "experience", we are not very good at it scientifically and as reasonably a total. Therefore, there are many approaches appeared, which aimed to formulate the decision-making methods for optimization of multi-objective problems. Many capable scholars have already dealt with the modeling aspects of this kind of problems from different viewpoints.

Inevitably, the kernel problem of every algorithm aims to provide a reasonable and scientific approach for decision-maker to conduct a judgment for a multi-objective problem with conflicting tendencies.

Here in this book, we will pay attention mainly to the probability-based approach for optimization of multi-objective problems of material selection, which is developed recently by our group in viewpoint of system theory rationally. As an accompanying description of the comprehensive statement, the features of other previous formal techniques which were formulated for optimization of multi-objective problems are summarized and reviewed in this chapter.

M. Zheng et al., *Probability-Based Multi-objective Optimization for Material Selection*, https://doi.org/10.1007/978-981-99-3939-8_2

2.2 Previous Approaches for Multi-objective Optimization of Material Selection

Presently, more than 160,000 engineering materials including metals and non-metals appeared in the world [1–3]. These large amounts of materials form a material library, which is available to designers and engineers along with many manufacturing processes and selection attributes (perhaps conflicting with each other). Since its complexity, it makes material selection a difficult task. Therefore, for the proper design and engineering application of materials, a general quantitative evaluation of material selection is required.

So far, a series of methods have been developed to analyze the large amount of data involved in the material selection process, which aims to obtain compromised results [2]. It includes the qualitative and quantitative methodologies roughly. The qualitative approach contains empirical method, analogy method, substitution method, and trial and error method. The quantitative method includes many approaches, which convert requirements of performance, manufacturing process, and economic benefits into quantitative performance indicators of materials for processing in quantitative manner.

2.2.1 Qualitative Approach

2.2.1.1 Empirical Method

In this method, the selection of materials was based on the successful experiences in the previous work concerning material applications and selections for the same parts, or the materials recommended by the design manual for such parts. If there is similar product at home and abroad, the materials used in the similar parts can be applied through technology introduction or material composition and performance testing accordingly.

2.2.1.2 Analogy Method

By referring to the material conditions of other types of products with similar functions or conditions of use and the actual utilities, the same or similar materials could be selected after reasonable analysis and comparison.

2.2.1.3 Substitution Method

In repairing mechanical parts or substituting parts, if the originally selected material is not available or not suitable for some reasons, the main performance indexes of the original material can be referred to, and another material with similar performance can be selected. In order to ensure the safety of the parts in application, the quality and performance of the substitute materials should be not lower than the original materials in general.

2.2.1.4 Trial and Error Method

If it is a key part of the new design, the material should be selected according to the whole process of certain material selection demands. If the material test results fail to meet the performance requirements of the design, the gap should be found, the reason should be analyzed, and the selected material grade or heat treatment method should be improved. Then one carries out the test until the results meet the requirements, and thus the selected materials and their heat treatment methods can be determined according to the results. The selected material can well meet the use and processing requirements of the part remains to be tested in practice. Therefore, the work of material selection not only runs through all stages of product development, design, manufacturing, etc., but also needs to find problems in time during use and continuously to improve materials.

2.2.2 Quantitative Approach

2.2.2.1 Farag Comprehensive Method

Farag et al. proposed a comprehensive method for the related activities of product design, material selection, and cost estimation [1].

First, the design constraints and performance requirements are used to narrow the wide range of engineering materials to a limited number of candidate materials. Each candidate material is used to develop the best design, which is then used for cost estimation. Optimization technology, namely benefit–cost analysis, is used to select the best design–material combination for preliminary material selection. This method lacks a quantitative comparison of other attributes, such as the difficulty of manufacturing and processing technology, and the environment. In his book, the performance index method is introduced [2], which is in fact an alternative form of simple additive weighting (SAW) that attempts to overcome the shortcomings of combining different units in the original weighted attribute method by introducing a scaling factor [2]. What is more serious is that during the scaling process, the scaled value of the beneficial material performance index is proportional to the normal material performance value, while the scaled value of the unbeneficial material performance index is inversely proportional to the normalized material performance value, which obviously sets the beneficial material performance value and unbeneficial attribute index in a non-equivalent or inconsistent position [2].

2.2.2.2 Analytic Hierarchy Process (AHP)

Saaty proposed the analytic hierarchy process (AHP) [4]. AHP is a measurement method through pairwise comparison, which relies on expert judgment to give priority. Through scaled processing, intangible assets are relatively measured [4]. The comparison is made using an absolute judgment scale, which represents the degree of one element dominating another element in terms of a given attribute. In fact, the normalization factor (denominator) is subjectively selected in the scaling process, which influences the exact value of each decision matrix element and the final competitive result significantly. Different scaling arithmetic operations, such as vector normalization, linear scaling, extreme value processing, and standard deviation normalization, result in quite different consequences.

2.2.2.3 VlŠekriterijumska Optimizacija I KOmpromisno Resenje (VIKOR)

Opricovic developed VlŠekriterijumska Optimizacija I KOmpromisno Resenje (VIKOR) method [5], the compromise ranking list, the compromise solution, and the weight stability intervals for preference stability of the compromise solution obtained with the initial (given) weights are all determined. The "closeness" to the "virtual ideal solution" is used to measure the multi-criteria ranking index Q. On the other hand, it introduces an additional artificial weighting factor ν in the assessment procedure for VIKOR value Q [5].

2.2.2.4 Technique of Ranking Preferences by Similarity to the Ideal Solution (TOPSIS)

Hwang and Yoon put forward technique of ranking preferences by similarity to the ideal solution (TOPSIS) preliminarily in 1981, and further development was conducted by Chen and Hwang in 1992 [6]. In TOPSIS method, there are two "virtual ideal points", i.e., the so-called a positive ideal solution and a negative ideal solution. TOPSIS employed the method of maximizing the distance to the negative ideal solution and minimizing the distance to the ideal positive solution to obtain the best alternative. In addition, Euclidean distances and normalized decision matrix are employed in TOPSIS method to conduct the assessment of the alternatives to their negative ideal solution and positive ideal solution. The commonly used normalization methods for TOPSIS are given in Table 2.1, and the distance measures (functions) for TOPSIS are listed in Table 2.2.

By ranking the Euclidean distances, the preference order of alternatives is gotten. However, the validity of the normalized factor and the "virtual ideal points" in the normalization are not rationally justified.

Table 2.1 Some normalization methods in TOPSIS

Name	Method
1. Vector normalization	$r_{ij} = \frac{x_{ij}}{[\sum_{i=1}^{m} x_{ij}^2]^{0.5}}$, $i = 1, 2, \ldots m; j = 1, 2, \ldots n$
2. Linear normalization-1	$r_{ij} = x_{ij}/x_j{*}$, $i = 1, 2, \ldots m; j = 1, 2, \ldots n$; $x_j{*} = \max_i \{x_{ij}\}$ for beneficial attributes $r_{ij} = x_j'/x_{ij}$, or $r_{ij} = 1 - x_{ij}/x_j{*}$, $i = 1, 2, \ldots m; j = 1, 2, \ldots n$ $x_j' = \min_i \{x_{ij}\}$ for unbeneficial attributes
3. Linear normalization-2	$r_{ij} = (x_{ij} - x_j')/(x_j{*} - x_j')$ for beneficial attributes $r_{ij} = (x_j{*} - x_{ij})/(x_j{*} - x_j')$ for unbeneficial attributes
4. Linear normalization-3	$r_{ij} = \frac{x_{ij}}{\sum_{i=1}^{m} x_{ij}}$, $i = 1, 2, \ldots m; j = 1, 2, \ldots n$
5. Non-monotonic normalization	$\exp\left(\frac{-z^2}{2}\right)$, $z = \frac{x_{ij} - x_j^0}{\sigma_j}$, $x_j{}^0$ expresses the most preferable value, σ_j indicates the standard deviation of alternative ratings with respective to the j-th attribute

Table 2.2 Distance measures (functions) for TOPSIS

Name	Method
1. Minkowski L_p metrics	$L_p(x, y) = \left\{\left(\sum_{j=1}^{n} \lvert x_j - y_j \rvert\right)^p\right\}^{1/p}$, n is dimensional or direction number, $p \geq 1$
2. Weighted L_p metrics	$L_p(x, y) = \left\{\left(w_j \sum_{j=1}^{n} \lvert x_j - y_j \rvert\right)^p\right\}^{1/p}$, n is dimensional number, $p \geq 1, 2, 3\ldots$, w_j is the weight of j-th dimension or direction

2.2.2.5 Multi-objective Optimization on the Basis of Ratio Analysis (MOORA)

Brauers, el al proposed multi-objective optimization (MOO) on the basis of ratio analysis (MOORA) for discrete alternatives [7]. In this method, a ratio system is in need, and each response of an alternative is divided by a normalized factor (denominator), which is representative of that objective for all alternatives. Furthermore, these responses are added or subtracted according to the case of maximization or minimization in the optimization. Obviously, in MOORA method, the reasonability of the selection for the normalized factor to each attribute of an alternative and the algorithm of "adding or subtracting according to the case of maximization or minimization in the optimization" in the arithmetic operation process are not clear.

2.2.2.6 M. F. Ashby's Method

Ashby developed a material selection chart for various materials. Two performance indicators are presented in the chart [3]. Various physical properties of the materials in the chart (i.e., electrical conductivity, elastic modulus, etc.) are empirically related to fundamental parameters (i.e., density, heat capacity, etc.). Ashby's method is suitable for preliminary screening of materials only since other indicators are short for effective comparisons in this method, such as processing technology and difficulty of manufacturing and environment. Therefore, this method is suitable for initial screening of materials.

In the following, some examples for initial screening of materials are given.

Case 1 Cantilever beam with constant cross section

Taking a cantilever beam with constant cross section as an example, the problem is to conduct the analysis of material selection without elastic deformation failure, or prevent excessive elastic deformation of the cantilever beam. As shown in Fig. 2.1, the length of the arm beam is l, the section of the square is with the width of a, the applied load is F, and the allowable deflection is $[y]$.

The analysis for material selection problem is as follows:

The maximum deflection of the beam is $y_{\max} = \frac{4l^3F}{Ea^4}$;
The stiffness condition of the beam is $y_{\max} \leq [y]$;
From this, we get $a \geq \left(\frac{4l^3F}{[y]}\right)^{1/4} \cdot \left(\frac{1}{E}\right)^{1/4}$.

It can be seen that when other conditions remain unchanged, in order to minimize the cross-sectional size of the beam, the material with the highest E value should be selected. In addition, the mass M of the beam is $M = la^2\rho$, where ρ is the material density. Substituting in the expression of a, it gets $M \geq \left(\frac{4l^5F}{[y]}\right)^{1/2} \cdot \left(\frac{\rho^2}{E}\right)^{1/2}$.

It can be seen that when the mass of the beam is required to be as light as possible, the smallest value of $(\rho^2/E)^{1/2}$ for material should be selected. Table 2.3 shows performance parameters of several materials.

Notice: In general, elastic modulus mainly depends on the nature of the material, and it is not sensitive to changes of microstructure and composition. Therefore, when selecting materials with stiffness as the main index, cheap low carbon steel or cast iron should be considered. For metal structures such as bridges, low carbon steel or ordinary low alloy steel can be used; various types of cast iron can be used for frame and bed.

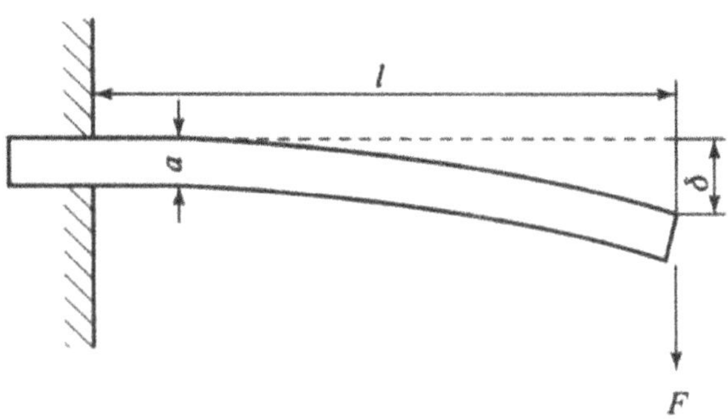

Fig. 2.1 Material selection of cantilever beam with equal section

Table 2.3 Performance parameters of several materials

Materials	Elastic modulus E (GPa)	Density ρ ($\times 10^3$ kg/m^3)	Price C (¥RMB / ton)	$(\rho^2/E)^{1/2}$ ($\times 10^{-2} N^{1/2} m^{-2}$)	$C \cdot (\rho^2/E)^{1/2}$ ($\times 10^{-4} N^{-1/2} m^{-2}$ ¥RMB)
Concrete	48	2.5	580	11.4	66.1
Wood	12.5	0.6	860	5.4	46.4
Steel	210	7.8	900	17.1	153.9
Aluminum alloy	73	2.7	4600	10.1	464.6
Fiberglass composite	30	1.8	6600	10.2	673.2
Carbon fiber composite	150	1.5	400,000	3.8	15,200
Plexiglass	3.4	1.2	10,000	20.5	2050

Case 2 Leaf spring

Taking a leaf spring as the next example, the consideration of yield strength in material selection to prevent plastic deformation failure is analyzed, which is to prevent excessive elastic deformation. Let the length of the leaf spring be l, the width be b, and the thickness be t, as shown in Fig. 2.2.

The force of the leaf spring is equivalent to the middle loaded support beam. If the self-weight is omitted, its deflection is $y = \frac{Fl^3}{4Ebt^3}$.

Figure 2.3 shows the stress distribution of the leaf spring section. The stress at the center line is 0, the stress at the surface is the largest, and its value is $\sigma = \frac{3Fl}{2bt^2}$.

The leaf spring is not allowed to undergo plastic deformation during work, so its working stress is required to be less than the yield strength, namely: $\sigma = \frac{3Fl}{2bt^2} < R_{el}$. From this condition, it can be obtained $R_{el}/E > 6yt/l^2$. In the above formula, R_{el}/E is related to the material, and $6yt/l^2$ is the design required quantity. Obviously, when the selected material meets the requirements of the above formula, it can ensure that the leaf spring does not undergo plastic deformation. Table 2.4 gives the values for the materials that can be used to make the springs.

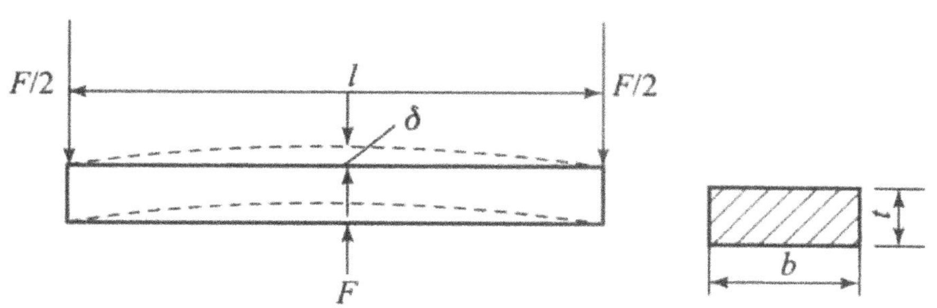

Fig. 2.2 Material selection of leaf springs

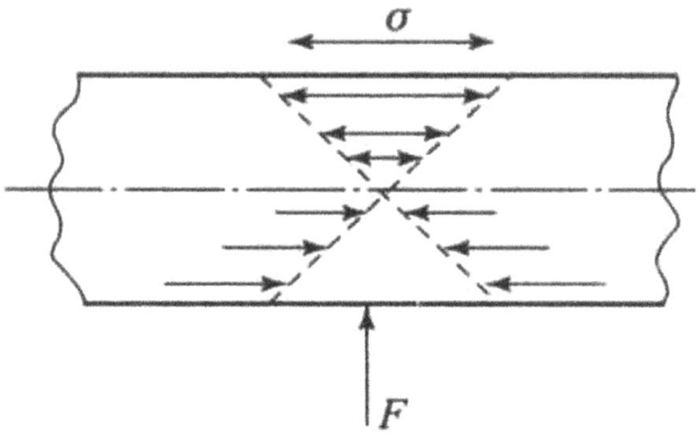

Fig. 2.3 Stress distribution of leaf spring section

Table 2.4 Values R_{el}/E of materials that can be used to make springs

Materials	E (GPa)	R_{el} (MPa)	R_{el}/E ($\times 10^{-3}$)
Cold rolled brass	120	638	5.32
Cold rolled bronze	120	640	5.33
Phosphor bronze	120	770	6.43
Beryllium bronze	120	1380	11.5
Spring steel	200	1300	6.5
Cold rolled stainless steel	200	1000	5.0
Nimonic superalloy	200	614	3.08

2.2.3 Discussion and Summary of the Previous Approaches for Multi-objective Optimization of Material Selection

From above discussion, the normalization is an indispensable process in the above "additive" algorithms to allow diverse criteria to form an "unique target" in eliminating the difference of dimensional units, the introductions of artificial factors or subjective denominator in the scaling process for the normalization of decision matrix in some methods are involved, and the final result depended on the normalization process significantly [8]. Different normalization methods will produce considerable differences in the result. Askoldas Podviezko and Valentinas Podvezko showed that different types of transformation and normalization of data applied to popular MOO methods such as SAW or TOPSIS produced considerable differences in the evaluation. Consequently, attention has to be paid to make a choice of the type of normalization. The actual problem is that it is unknown which normalization method is better and how to decide final result of material selection in consideration of the individual results from TOPSIS with different normalization methods [8]. Therefore, above algorithms could not be seen as complete quantitative ones, which are at most semiquantitative approaches in some sense.

Besides, the treatment for beneficial performance index and unbeneficial performance index is non-equivalent or inconsistent in some algorithms. More seriously, the multi-objective optimization method proposed so far used the "additive" algorithm for the normalized evaluation index. Its inherent shortcoming is that it takes the form of "union" from the spirit of set theory. In fact, for the evaluation of "simultaneous optimization of multi-performance utility index", the form of "intersection" in set theory and "joint probability" in probability theory should be more suitable to be adopted [9–12]. Therefore, comprehensive study for materials selection is still in need so that a quantitative and overall approach can be developed.

In response to the requirement of "simultaneous optimization of multiple indexes", the multi-object optimization is reformulated from the viewpoint of system theory and spirits of set theory and probability theory [9–12]. In the following chapters of this book, the probability-based method for multi-object optimization for material selection is developed on basis of probability theory as an overall consideration of the "simultaneous optimization of multiple indexes". The new idea of preferable probability is introduced, each utility index of the candidate scheme contributes a partial preferable probability quantitatively, and the overall/total preferable probability of a candidate scheme is the product of the partial preferable probabilities of all possible utility indexes of material performance indicators, which is the integrated operation of all possible material performance indicators for the candidate material in the viewpoint of system theory. The total preferable probability of a candidate material is the unique and decisive index for the material selection process.

2.3 Fundamental Consideration of Multiple objective Optimization for Material Selection

2.3.1 Statement of Situation

In facing material library with a huge number of materials, designers must comprehensively consider the specific manufacturing process on equipment and the basic properties of the material in the material selection process in order to select suitable materials reasonably. However, the diversity of production processes and the functionality of process equipment lead to the complexity of material selection, making material selection one of the important links in equipment design and manufacturing. Therefore, in order to facilitate the rational design and application of engineering materials, it is particularly important to carry out quantitative evaluation in the process of material selection.

2.3.2 Basic Principles for Selection of Equipment Materials

The selected materials for parts should be suitable for their processing methods. Material selection is the first issue to be considered in product design. There are many materials and processing methods, so material selection is often a complex and difficult job concerning judgment and optimization process. When selecting materials and forming processes, we must first consider whether the material properties for the parts can meet the requirements under their servicing conditions, and thereafter the consideration is whether the forming process of the parts is easy to conduct or applicable. At the same time, it is also necessary to consider whether its production and use are economical. Therefore, when selecting materials and forming processes, one thing is to meet the performance requirements; the other thing is to meet the processing and manufacturing requirements; the third issue is to have high economic benefits.

Following factors and procedures are involved in material selection with quantitative assessment in general:

A. Convert the performance requirements of parts into quantitative performance indicators and their utilities of materials;
B. Convert the manufacturing process requirements of parts into quantified index parameters and their utilities;
C. Convert the economic benefit requirements of parts into quantified index parameters and their utilities.

2.3.3 Basic Procedure for Material Selection

The basic idea of material selection is shown in Fig. 2.4. First, according to the principle of service performance, the analysis of the working conditions and failure modes of the process equipment is conducted preliminarily; then service performance that the component has to withstand is determined. Afterward, the service condition of the component is converted into the service performance of the material indicators, the preselected materials can be determined from consulting relevant manuals. Generally, the preselected materials are not unique. The performance, process performance, and economy of the preselected materials are comprehensively analyzed to determine the selected materials.

The principles of material selection are described as follows.

1. ***Applicable principle of material performance***

The service performance mainly refers to the performance that the component should have in the use state, including the mechanical properties, corrosion resistance, and physical properties of the material. The material performance principle is the starting point for material selection. Under defined environment, certain function of a component must be guaranteed by using the performance principle. When the selected material has sufficient performance for service, the next considerations will focus on their performances of processing and economy. The performance requirements for material are put forward generally on the basis of the analysis of its working conditions and failure modes of the components.

The working conditions of the components can be roughly summarized as follows.

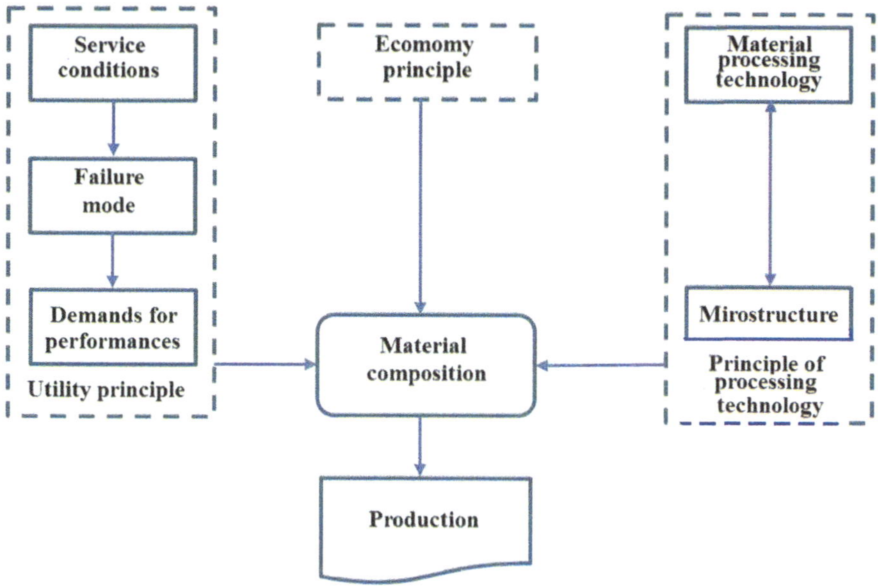

Fig. 2.4 Basic factors of material selection

A. The stress conditions of components involve load type and form, such as the type of load (static load, alternating load, impact load, etc.), the form of load (tensile, compression, torsion, bending, shear, etc.), the size and distribution of the load (uniform load, etc.), distribution or large local stress concentration, and so on.
B. The working environment of components is mainly concerning temperature and medium conditions. Temperature conditions contain different temperature range, such as low temperature, normal temperature, high temperature, or variable temperature and medium conditions such as corrosion, nuclear radiation, fouling, or friction.
C. Special requirements of components indicate non-mechanical factors, such as fast heat transfer, anti-vibration, and lightweight.

Components working under certain working conditions often fail due to various reasons. In order to prevent certain modes of possible failure, it is necessary to put forward certain requirements on the performance indicators of materials during design. Table 2.5 lists the failure modes, working modes, and main performance indicators of several components.

2. ***Principle of processing technology performance***

The technological properties of a material represent the ease with which the material is processed. Any part of the selected material is manufactured through a certain processing technology. Therefore, the quality of the material process performance will directly affect the quality, production efficiency, and cost of the part. The

Table 2.5 Failure mode, working mode, and main performance index of several components

Part	Working condition	Main failure mode	Main mechanical performance index
Bolts	Alternating tensile stress	Fracture due to excessive plastic deformation or fatigue	Yield strength, fatigue strength, hardness (HBW)
Transmission gears	Alternating bending stress, alternating contact pressure, impact load, and tooth flank friction	Teeth fracture, excessive wear, or fretting fatigue	Bending strength, fatigue strength, contact fatigue strength, hardness (HRC)
Shaft	Alternating bending stress, impact load and neck friction, torsional stress	Fatigue fracture	Yield strength, fatigue strength, hardness (HRC)
Spring	Alternating stress, vibration	Loss of elasticity, fatigue fracture	Elastic limit, yield ratio, fatigue strength
Rolling bearings	Alternating compressive stress under point or line contact, rolling friction	Excessive wear, fatigue damage	Compressive strength, fatigue strength, hardness (HRC)

selected material should have good process performance, that is, simple process, easy processing and forming, low energy consumption, high material utilization rate, and good product quality.

The technological properties of metal materials mainly include casting properties, pressure processing properties, welding properties, machining properties, and heat treatment properties.

A. Casting properties. The casting properties of materials are generally evaluated comprehensively according to their fluidity, shrinkage characteristics, and segregation tendency.
B. Press processing performance. There are many types of press working, which can be roughly divided into two categories: hot working, mainly hot forging, hot extrusion, etc.; cold working, mainly cold stamping, cold pier, cold extrusion, etc. When selecting materials, the components with large loads and complex forces (important shafts, internal combustion engine connecting rods, gearbox gears, etc.) should choose medium and low carbon steel or alloy structural steel, forged aluminum, and other materials with good forgeability. Forging is carried out, and necessary heat treatment is conducted to strengthen the structure and improve the mechanical properties. Many light industrial products (such as bicycles, metal parts on household appliances) generally have a small load, but require beautiful color, lightweight and large batches, surface protection, and decorative treatment.

C. Welding performance. In the machinery industry, the main objects of welding are various steels. Weldability can be roughly evaluated by carbon equivalent. When the carbon equivalent exceeds 0.44%, the weldability of the steel is extremely poor. Therefore, the higher the carbon content of the steel and the higher the alloying element content, the worse the weldability. Steel with too high carbon equivalent should not use welding forming method to make part blank. Many containers, pipelines, steam boilers, and other products as well as some engineering structures (generally larger in size, requiring good air tightness, and being able to withstand a certain pressure) should be welded with materials with good welding performance such as low carbon steel and low alloy steel. Aluminum alloys and titanium alloys are easily oxidized and need to be welded in a protective atmosphere, and their welding performance is not good.
D. Machinability. All kinds of machining (mainly cutting and grinding) are the most widely used metal processing methods in the industry. Most machine parts need to be cut, and materials with moderate hardness (170–230 HBW) and good cutting performance should be selected. The machinability of cutting aluminum and its alloys is good, while the machinability of austenitic stainless steel and high speed steel is poor. When the machinability of the material is poor, necessary heat treatment can be used to adjust its hardness or improve the cutting process to ensure the cutting quality.
E. Heat treatment process performance. Many metal components require heat treatment (especially quenching and tempering) to achieve the required mechanical properties. Therefore, the process performance of heat treatment, especially the hardenability, cannot be ignored when selecting materials. For parts that require overall hardening and large cross section, alloy steel with high hardenability should be selected; for workpieces with complex shapes and strict requirements for heat treatment deformation, alloy steel with high hardenability should also be used and slow cooling method to reduce quenching deformation. For workpieces that only require surface strengthening or simple shapes, materials with lower hardenability can be selected. When selecting materials, process performance is secondary to performance, but in some special cases, process performance may also become the main factor for material selection. Taking cutting as an example, under the condition of single-piece small batch production, the cutting performance of the material is not important. In mass production conditions, machinability becomes a decisive factor in material selection. For example, a factory once trial produced a 25 SiMnWV steel as a substitute for 20 CrMnTi steel. Although its mechanical properties are higher than 20 CrMnTi, it has high hardness after normalizing and poor machinability, which cannot be adapted to mass production, so it was not used.

3. ***Principles of economy***

On the premise of meeting the requirements of component service performance and processing technology performance, economy is also the major issue that must be considered. The economy of material selection is not only the price of the selected materials, but also the total cost of component production in entire lifetime. The material cost should be considered in combination with the processing and manufacturing, installation, operation, inspection, maintenance, environment impact, replacement, and equipment life of the components, and the total cost of financial management should be considered. It is also important to consider the availability of material sources and compliance with the country's resource policy when selecting materials.

In principle, the service condition, the processing technology, and the whole cost in its entire lifetime are needed to be considered by means of utilities of material performance indicators in material selection.

2.4 Conclusion

This chapter collected and reviewed the previous approaches for material selection with multi-objective optimization mainly and stated preliminarily the fundamental procedure for material selection subsequently. It reveals the complexity of material selection containing conflicting material performances (attributes) within one integral body (system).

References

1. M.M. Farag, E. El-Magd, An integrated approach to product design, materials selection and cost estimation. Mater. Des. **13**, 323–327 (1992)
2. M.M. Farag, *Materials and Process Selection for Engineering Design*, 4th edn. (CRC Press, Taylor & Francis Group, New York, 2021), pp. 328–334
3. M.F. Ashby, *Materials Selection in Mechanical Design* (Butterworth-Heinemann Ltd, Burlington, 1992)
4. A. Jahan, K.L. Edwards, A state-of-the—art survey on the influence of normalization techniques in ranking: Improving the materials selection process in engineering design. Mater. Des. **65**, 335–342 (2015)
5. S. Opricovic, G.H. Tzeng, Compromise solution by MCDM methods: a comparative analysis of VIKOR and TOPSIS. Eur. J. Oper. Res. **156**, 445–455 (2004)
6. P. Wang, Z. Zhu, Y. Wang, A novel hybrid MCDM model combining the SAW, TOPSIS and GRA methods based on experimental design. Inf. Sci. **345**, 27–45 (2016)
7. W.K.M. Brauers, E.K. Zavadskas, The MOORA method and its application to privatization in a transition economy. Control Cybern. **35**, 445–469 (2006)
8. W.C. Yang, S.H. Chon, C.M. Choe, J.Y. Yang, Materials selection method using TOPSIS with some popular normalization methods. Eng. Res. Express **3**, 015020 (2021)
9. M. Zheng, Y. Wang, H. Teng, A new "intersection" method for multi—objective optimization in material selection. Tehnički Glas. **15**(4), 562–568 (2021). https://doi.org/10.31803/tg-20210901142449
10. M. Zheng, Y. Wang, H. Teng, Hybrid of the "intersection" algorithm for multi-objective optimization with response surface methodology and its application. Tehnički Glas. **16**(4), 454–457 (2022). https://doi.org/10.31803/tg-20210930051227

11. M. Zheng, H. Teng, Y. Wang, Robust design in material machining process on basis of probability multi-objective optimization. Materialwiss. Werkstofftech. **54**(2), 180–185 (2023). https://doi.org/10.1002/mawe.202200162
12. M. Zheng, Y. Wang, H. Teng, A novel method based on probability theory for simultaneous optimization of multi-object orthogonal test design in material engineering. Kovove Mater. **60**(1), 45–53(2022). https://doi.org/10.31577/km.2022.1.45

Chapter 3
Fundamental Principle of Probability-Based Multi-objective Optimization and Applications

Abstract The inherent shortcomings of previously proposed multi-objective optimization methods are employing "additive" algorithm for the normalized evaluation index and weighting factor, which implies to take the form of "union" in the spirit of set theory. In fact, for the evaluation of "simultaneous optimization of multi-performance utility index", the form of "intersection" in set theory and "joint probability" in probability theory should be more suitable for the problem. The viewpoint of system theory is consistent with this understanding as well. In this chapter, the new idea of preferable probability is introduced to reflect the degree of preference of the candidate's utility in the selection of multi-objective optimization in viewpoint of system theory; all the utility indexes of candidate schemes are divided into two types, i.e., the beneficial type and the unbeneficial type for the selection of the schemes; each utility index of the candidate scheme contributes a partial preferable probability quantitatively, and the overall/total preferable probability of a candidate scheme is the product of all partial preferable probabilities in the spirit of probability theory, which thus transfers the multi-objective optimization problem into an overall (integrated) single-objective optimization issue naturally. The total preferable probability is the uniquely decisive indicator in the competitive selection process. In addition, examples of applications in material selection and some other businesses in broader and more general fields are given, and the results show the effectiveness of the new methodology.

3.1 Introduction

In current years, more than 160,000 useful materials are available to designers and engineers for their selection and use [1], which consists of a material library for customers. Besides, many complicated manufacturing processes are involved in practical material engineering in addition to the complex relationships among different material parameters, which makes the material selection a perplexed task.

So, it is necessary to have a quantitative approach for material selection so as to conduct an appropriate design and application of materials in engineering.

M. Zheng et al., *Probability-Based Multi-objective Optimization for Material Selection*, https://doi.org/10.1007/978-981-99-3939-8_3

It experienced about half a century since the early publication of some pioneer works for material selection [2–4], a series of algorithms for multi-objective optimization have been employed to the issue of material selection, such as Analytical Hierarchy Process (AHP), Technique of ranking Preferences by Similarity to the Ideal Solution (TOPSIS), VIšekriterijumska Optimizacija I KOmpromisno Resenje (VIKOR), multi-objective optimization on the basis of ratio analysis (MOORA), and Pareto optimality[1–5].

It is no doubt that material selection involves issues of "simultaneous optimization of multiple objectives" or "multiple criteria decision making" in principle. However, the previously proposed algorithms for multi-objective optimization or multi-criteria decision making (MCDM) so far adopted the "additive" algorithm and normalization/parameterization for the evaluation indexes, and some even include personal factors, such as that in TOPSIS, MOORA, and VIKOR's approaches [2–4]. The "additive" algorithm of multiple indexes in evaluation has its inherent characteristics, which is not coincident with the real intention of "simultaneous optimization of multiple indexes", but equivalent to taking the form of "union" in the spirit of set theory instead [6]. In fact, in the respect of probability theory, the essence of "simultaneous optimization of multiple indexes" has the meaning of "joint probability" of such multiple indexes as independent events analogically.

Besides, owing to the employment of personal factors, the relevant algorithms cannot be considered as full quantitative methods in some sense. On the other hand, the format of normalization and the selection of scaled factor (denominator) in their normalization process are all problematic and troublesome issues. It may lead to quite different consequences with different normalization algorithms [7]. Additionally, different types of normalization and transformation of data are employed for previously mentioned multi-criteria decision-making approaches such as simple additive weighted (SAW) algorithm or TOPSIS, which might induce sensible differences in the assessment. The alternatives adopted the "additive" algorithm as well, for example synthesis weighting algorithm to conduct the comparative assessments of subjective and objective characteristic of weightings [8–10]. Conclusively, the shortcomings of the weighted sum method manifest following respects: (1) the artificiality in determining the weighting and (2) the discrepancy between the actual algorithm and the essence of "simultaneous optimization of multiple indexes" [11].

Above situation indicates that the study for appropriate expression of "simultaneous optimization of multiple objectives" for issue of materials selection is still required, so that an overall quantitative consideration can be obtained.

In this chapter, a probability theory approach for material selection is developed by introducing a new concept of preferable probability, and the overall/total preferable probability of a candidate material is the integral consideration of all possible material performance indicators of the candidate material in comprehensive respects. The overall preferable probability of a candidate material is the unique decisive index for the material to gain the competition reasonably and impersonally.

3.2 Multi-objective Optimization in Viewpoint of System Theory

Multi-objective optimization means that the object (system) contains multiple objectives, which cannot be separated and need to be optimized at the same time. This is its integrity. For example, the materials used to make airplanes need to be strong and tough, but also light and tolerant to the environment. The attributes (targets) such as strength, toughness, lightness, and environmental tolerance are inseparable indicators contained in a material, and there is no way to separate any of them and exist alone. In other words, a material wants to have high strength and lightweight (low density), but high strength and lightweight (low density) are contradictory indicators, and they exist in this material at the same time and are inseparable. The optimization problem is that these attributes must be "optimized simultaneously" to make the system (organization) play its due role! The attributes are also interrelated, so it is impossible to adjust only one of them without affecting the others! The environmental adaptability of the system is also an important performance index to determine its reliability and success or failure. Therefore, the multi-objective optimization problem is the overall optimization problem from the viewpoint of system theory. Every goal is an organic part of the system. Dealing with this multi-objective optimization problem of "simultaneous optimization" is to optimize the problem as a whole, that is, to make the system in the best state as a whole, and all parts of it are "united" under the banner of "overall optimization" and "working" in coordination. Only when all the parts (targets) in the system are coordinated with each other, can the comprehensive optimization of the system objectives and the optimization of the functions of the whole system be realized. Therefore, the optimization principles of system theory include (1) overall optimization; (2) graded optimization in each stage; and (3) give consideration to all parties [12]. According to the viewpoint of system theory, "the whole system is not equal to the simple sum of its components", while "the whole is greater than the sum of its parts". However, in the previous approaches for multi-objective optimization, the method of "linear weighting" or "ε-constraint" is adopted, in which "adding weights" or "selecting one of the k objectives as the optimization objective and transforming the remaining $(k-1)$ objectives into constraints" is contrary to the connotation of "simultaneous optimization of multiple objectives" undoubtedly and away from the intrinsic essence of "overall optimization of the system". In addition, the selections of weighting factors and normalization factors in the "linear weighting" method are also problematic. Because the intrinsic connotation of "multi-objective optimization" has the meaning of "multi-objective optimization at the same time", it is "overall optimization of the system" from the point of view of system theory. Therefore, it is necessary to find the "intersection" among the objectives to make them "coordinate" with each other, so as to realize the optimization of the overall function of the system. The concept of "intersection" comes from "set theory", which relates to two sets A and B, the set consisting of all elements belonging to both set A and set B is called the intersection of set A and set B, and is recorded as $A \cap B$. In "probability theory", the probability $P(A \cap B)$ of two independent events "appearing at the same time" is equal to $P(A) \cdot P(B)$, that is, $P(A \cap B) = P(A) \cdot P(B)$, which is called the joint probability of two independent events A and B [13].

Furthermore, according to the viewpoint of system theory, the system can have various forms, and each component in the system can also have various forms. The materials in multi-objective material selection have the form of solid system, while the responses, such as elastic modulus, tensile strength, and elongation of materials, are more similar to the form of conceptual system. When we regard each objective (attribute) in multi-objective optimization as an event, the problem of "simultaneous optimization of multiple objectives" becomes a probability problem of "simultaneous occurrence of multiple events". Moreover, if the events are independent, the joint probability of the "overall optimization of the system" problem of "multiple events occurring at the same time" is equal to the overall probability P_t, $P_t = P_1 \cdot P_2 \cdot \ldots \cdot P_j \cdot \ldots = \prod_{j=1}^{m} P_j$. In this formula, P_j represents the probability of the j-th target and m is the total number of targets. In this way, it thus transforms a multi-objective optimization problem into an equivalent probability problem. For the simultaneous optimization of multiple objectives, Derringer et al. and Jorge et al. once put forward the concept of a satisfaction function [14, 15], which converts the response value of each objective into a satisfaction value and then combines all the satisfaction values by using the geometric average method to get a total satisfaction value to represent the overall evaluation of the combined response. However, this method simply does not conform to the intrinsic connotation of multi-objective simultaneous optimization from the perspective of probability theory.

3.3 Arithmetic of Probability Treatment

The exploration probability has been a branch of mathematics that can be back to more than 300 years ago to discuss questions corresponding to games of chance. The attraction of probability theory has been continuously increasing due to the needs of decision-making problems in science and engineering involving likelihood and uncertainty [6].

The understanding of its nature and sources of uncertainty is the basis for the treatment to the problem; thereafter, proper mathematical models can be employed to deal with it. Uncertainty means mainly unclearness and fuzziness in engineering by conventional measurement for usual physical parameters or variables. Probability theory might be an appropriate methodology to deal with the things with uncertainty [6].

Probability and reliability have the function to measure the likelihood, and characterization of likelihood has unique properties and treatments in mathematics, which are briefly collected in this section for our use.

1. **Sets**

A set is defined as a collection of elements or components, such as $\Omega = \{A, B, X, Y\}$ and $\Psi = (p, q: p > 0, q < 3)$. A subset is one part of a set.

2. **Events**

The set of all possible outcomes of an experiment (or a system) consists of the sample space S. A sample space constitutes of points corresponding to all possible outcomes.

Each outcome for the system is a unique element in the sample space. An *event* is a subset of the sample space. An event with no sample points is an empty set or called the *impossible event*. A set including all the sample points is named the *certain event S*. The certain event equals to the sample space, which is called the universal set.

3. Union and joint

For the subsets A and B attributing to set Ω, their union is denoted by $A \cup B$, and their intersection is indicated by $A \cap B$ [6]. The symbol $\overline{A}$ denotes the complement of A in Ω. The notation of $A + B$ (the sum of A and B) expresses the conventional meaning that A and B are disjoint, in which case it represents the union of A and B, i.e., $A \cup B$. While, the notation of $\sum_{k=1}^{\infty} A_k$ is used for $\bigcup_{k=1}^{\infty} A_k$ only when the $A_k s$ are pairwise disjoint. The notation $A - B$ is used only if $B \subseteq A$, and it stands for $A \cap \overline{B}$. In particular, if $B \subseteq A$, then $A = B + (A - B)$.

The symmetric difference of A and B, that is, the set $(A \cup B) - A \cap B$, is denoted by $A \Delta B$.

The indicator function of the subset A is the function l_A: $\Omega \to \{0,1\}$ defined by

$$l_A(\omega) = \begin{cases} 1, & if\ \omega \in A, \\ 0, & if\ \omega \notin A. \end{cases}$$

Random phenomena are observed by means of experiments (performed either by man or nature). Every experiment leads to an outcome. The sample space Ω indicates the collection of all possible outcomes ω. Any subset A of the sample space Ω will be regarded as a representation of some event for the time being.

The outcome ω realizes event A if $\omega \in A$, if ω does not realize A, it realizes $\overline{A}$. The event $A \cap B$ is realized by the outcome ω if and only if ω realizes both A and B. Analogically, $A \cup B$ is realized by ω if and only if at least one event among A and B is realized. Two events A and B are called incompatible when $A \cap B = \emptyset$. In other words, event $A \cap B$ means no outcome, ω can realize both A and B, and event $A \cap B$ is impossible. The symbol $\emptyset$ refers to the empty set of an impossible event. Of course, Ω is called the certain event. The impossible event is the complement of the certain event Ω.

Considering again that the notation $\sum_{k=1}^{\infty} A_k$ is used for $\bigcup_{k=1}^{\infty} A_k$ only when the subsets A_k are pair-wise disjoint. In the term of sets, the sets $A_1, A_2, \ldots$, form a partition of Ω if $\sum_{k=1}^{\infty} A_k = \Omega$.

If $B \subseteq A$, event B implies event A, since ω realizes A when it realizes B. Each event is assigned a number in probability theory, the probability of the event.

4. Probability of Event

The probability $P(A)$ of an event $A \in F$ indicates the likeliness of its occurring. Since a function is defined on F, the probability P is required to satisfy a few demands, the axioms of probability.

A probability on (Ω, F) is a mapping P: $F \to R$ that meets.

(i) $0 \leq P(A) \leq 1$ for all $A \in F$;
(ii) $P(\Omega) = 1$; and.
(iii) $P\left(\sum_{k=1}^{\infty} A_k\right) = \sum_{k=1}^{\infty} P(A_k)$ for all sequences $\{A_k\}_{k \geq 1}$ in case of pairwise disjoint events.

Besides, $P(\overline{A}) = 1 - P(A)$, and $P(\emptyset) = 0$. Probability is monotone, if $B \subseteq A$ for any events A and B, then $P(A) \leq P(B)$.

5. Independence and conditioning

As two events A and B are called independent if and only if $P(A \cap B) = P(A) \cdot P(B)$.

There is a family $\{A_n\}_{n \in N}$ of events, which is called independent if for any finite set of indices $i_1 < i_2 < \cdots < i_k$ where $i_r \in N$ $(1 \leq r \leq k)$, $P(A_{i1} \cap A_{i2} \cap \cdots \cap A_{ik}) = P(A_{i1}) \times P(A_{i2}) \times \cdots \times P(A_{ik})$, it can be also said that the $A_m s$ $(m \in N)$ are jointly independent.

6. Independent variables

If there is $P(X = i, Y = j) = P(X = i) \cdot P(Y = j)$, $(i, j \in E)$, for two discrete random variables X and Y, then they are called independent.

7. Mean and variance

If x is a random variable with $E[|X|] < \infty$, x is integrable. In this case (and only in this case), the mean μ of x is defined as $\mu \equiv E[x] = \sum_{n=0}^{\infty} nP(x = n)$. If x is a square-integrable random variable, its variance is defined as $\sigma^2 \equiv E[(x - \mu)^2] = \sum_{n=0}^{\infty} (n - \mu)^2 \cdot P(x = n)$.

If A is some even, then the expectation of the indicator random variable $x = l_A$ is $E[l_A] = P(A)$.

8. Independence and the product

Extension of two independent events A and B, i.e., $P(A \cap B) = P(A) \cdot P(B)$, leads to the more general form for a family $\{A_i\}_{i \in I}$ events being assumed independent, where I is an arbitrary index, if for every finite subset $J \in I$, $P(\underset{j \in J}{\cap} A_j) = \prod_{j \in J} P(A_j)$.

9. Intersection and union of C and D in Venn diagrams

Figure 3.1 shows representation of the union of C and D in Venn diagram with shaded only. Figure 3.2 gives representation of the intersection of C and D in Venn diagram with shaded and dotted lines. Obviously, the meanings of union and intersection are with distinct difference.

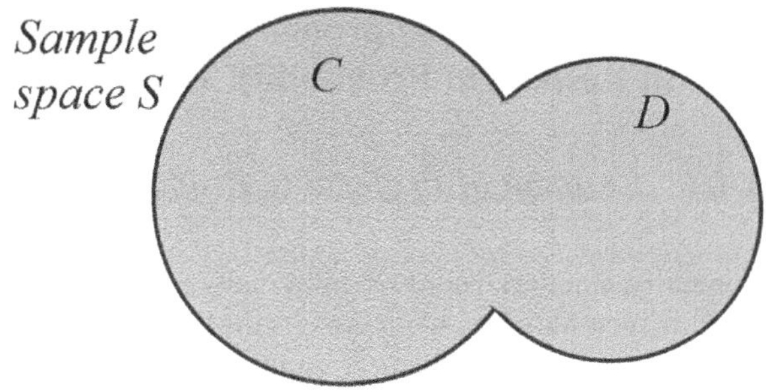

Fig. 3.1 Representation of the union of C and D in Venn diagram with shaded only

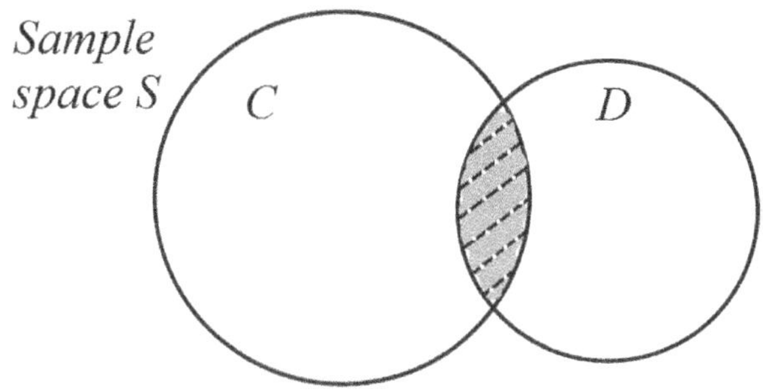

Fig. 3.2 Representation of the intersection of C and D in Venn diagram with shaded and dotted lines

10. Reliability

The reliability of a part is defined as the probability that the part meets some specific demands in determined servicing environments and conditions. Such as a beam, its reliability is defined as the probability that the ultimate moment capacity (strength) is greater than the applied total loadings so that it could withstand the loadings. The reliability R_e can be expressed by a mathematical formula as $R_e = P(R > L)$, where R expresses the strength or structural resistance of the part and L represents the applied total loadings which are expressed in the same units as that of strength R. Accordingly, the failure probability of P_f of the part is written as $P_f = P(R < L)$ $= 1 - R_e$.

11. Reliability of system in series

The reliability of a system in series relies on the level of correlations among the dormant failure events of the parts $S_\alpha(\alpha = 1, 2, \ldots, n)$. The residual of the system $(\overline{S})$ can be expressed as $\overline{S} = \overline{S_1} \cap \overline{S_2} \cap \overline{S_3} \cap \cdots \cap \overline{S_{n-1}} \cap \overline{S_n}$, where $\overline{S_\alpha}$ is the residual event of part α. As to independent failure events of the parts, the failure probability of the system can be assessed by $P(S) = 1 - P(\overline{S})$.

3.4 Quantitative Approach for Material Selection in Respect to Probability Theory

3.4.1 Concept of Preferable Probability

Generally, a material could exhibit many features in different respects; each material performance indicator expresses its one aspect in a sense.

As to the material selection, some of material performance indicator may possess the feature of "the higher the better", but other indicators have the feature of "the lower the better"; the former could be ascribed to the beneficial type, while the latter is imputed to the unbeneficial type in the process of material selection.

A real material as a whole is the organically integral body with both beneficial and unbeneficial types of performance indicators. It is inevitable for a material to have only beneficial or unbeneficial types of performance indicators for the material selection usually. Therefore, an overall consideration of the "simultaneous optimization of multi-objective" for material selection is needed in the respect of unbiased analysis, which makes the material selection really an overall and systemic task. Thus, both the beneficial and unbeneficial types of performance indicators must be appropriately treated following the rule of "simultaneous optimization of multi-objective" in the viewpoint of system theory, such that an authentic assessment could be obtained quantitatively.

In the spirit of probability theory, the probability of the entire event of "simultaneous optimization of multi-objective" corresponds to the product of the probability of each individual objective (event). Therefore, the common term "the higher the better" for the utility index of material performance indicator should be expressed quantitatively in terms of probability theory, which prompts us to seek the appropriate description in probability theory itself. The consequence in expressing the thought of "the higher the better" of the utility index of material performance indicator quantitatively is the introduction of a new concept of "preferable probability", which is used to reflect the preference degree of the utility index of material performance indicator of the candidate in the selection; i.e., it uses the term "preferable probability" to quantitatively represent the preference degree of the utility index of a material performance indicator in the material selection.

Moreover, it should formulate an actual expression for "preferable probability" from its utility index of material performance indicator. From the principle of simplicity, the direct and convenient assumption is that the partial preferable probability of the utility index with the character of "the higher the better" (beneficial index) in the material selection process is positively correlated with the value of the corresponding utility index in linear manner, i.e.,

$$P_{ij} \propto U_{ij}, \quad P_{ij} = \alpha_j U_{ij}, \quad i = 1, 2, \ldots, n, \; j = 1, 2, \ldots, m. \tag{3.1}$$

In Eq. (3.1), U_{ij} expresses the utility index value of the j-th material indicator of the i-th candidate material; P_{ij} indicates the partial preferable probability of the beneficial type of performance utility indicator U_{ij}; n shows the total number of candidate materials in the material group involved; m reflects the total number of the utility of performance indicators of each candidate material in the group; α_j represents the normalized factor of the j-th beneficial type of performance indicator. Of course, other complex form of expression for "preferable probability" from its utility index of material performance indicator can also be set up provided its physical meaning is realized.

Thereafter, in accordance with the general principle of normalization in probability theory [6], the summation of each P_{ij} for the index i in j-th utility of material property indicator is normalized to 1, i.e., $\sum_{i=1}^{n} P_{ij} = 1$, which leads to following consequence,

$$\sum_{i=1}^{n} \alpha_j U_{ij} = \sum_{i=1}^{n} P_{ij} = 1, \quad \alpha_j = 1/(n\overline{U_j}). \tag{3.2}$$

$\overline{U_j}$ expresses the arithmetic average value of the j-th utility index of material indicator in the material group to be assessed; α_j represents the normalized indicator of the j-th beneficial type of performance utility indicator.

Equivalently, the partial preferable probability of the unbeneficial type of performance utility indicator U_{ij} to the candidate material is negatively correlated with its utility value of the material indicator in linear manner, i.e.,

$$P_{ij} \infty (U_{j\,\max} + U_{j\,\min} - U_{ij}), \quad P_{ij} = \beta_j (U_{j\,\max} + U_{j\,\min} - U_{ij}),$$
$$i = 1, 2, \ldots, n, \ j = 1, 2, \ldots, m. \tag{3.3}$$

In Eq. (3.3), $U_{j\min}$ and $U_{j\max}$ express the minimum and maximum values of the utility index U_j of the material indicator in the material group, individually; β_j indicates the normalized indicator of the j-th unbeneficial type of performance utility indicator.

Analogically, according to the general principle of normalization of probability theory [6], one obtains

$$\beta_j = 1/[n(U_{j\,\max} + U_{j\,\min}) - n\overline{U_j}]. \tag{3.4}$$

Evidently, Eqs. (3.1) and (3.3) set the beneficial and unbeneficial utility of material indicator an equivalent position.

3.4.2 Probability-Based Approach

Following the spirit of "simultaneous optimization of multiple indexes" of probability theory, the overall/total preferable probability of the i-th candidate material is the product of its all partial preferable probabilities P_{ij} [6], i.e.,

$$P_i = P_{i1} \cdot P_{i2} \cdots P_{im} = \prod_{j=1}^{m} P_{ij}. \tag{3.5}$$

Overtly, the overall/total preferable probability P_i of a candidate material is the unique and decisive index for the material in the material selection process, ranking of all candidate materials can be conducted in sequence by their total preferable probabilities, and the selection of material will be performed according to the sequence as an overall consideration.

Thus, the overall/total preferable probability P_i of a candidate material transfers the problem of "simultaneous optimization of multiple indexes" into an overall (integrated) "optimization of single index" one.

So far, the new idea of preferable probability for material selection and its evaluation procedure have been formulized quantitatively.

3.5 Applications of the Probability-Based Method for Multi-objective Optimization in Material Selection

1. Material Selection for a Cryogenic Tank Design

According to Farag [1], an optimum material is required for a large cryogenic storage tank for transporting liquid nitrogen gas at temperature of -196 °C, materials with FCC structure will be the possible candidates due to their potential ductility at low temperatures, while many other metallic materials and plastics are excluded on this basis. Good weldability, mechanical properties, adequate toughness, high yield strength, and stiffness are all important and required. Lower density, lower specific heat, and lower thermal expansion coefficient are welcomed. By using these requirements, the quick analysis results in seven materials as the candidate materials for this cryogenic tank design. Table 3.1 presents the eight material performance indicators of the seven candidate materials for the cryogenic tank design [2]. Above material performance indicators could be taken as the utility indexes in this selection directly.

It can be seen from Table 3.1 that the first three performance indicators are beneficial type of indicators to the material selection, which has the characteristic of the higher the better, while the latter five performance indicators are unbeneficial type of indicators to the material selection, which has the characteristic of the lower the better.

All the candidate materials listed in Table 3.1 form a candidate material group for the selection.

Furthermore, it assumes that each material indicator contributes its effort to a partial preferable probability of the candidate material individually. According to the statement in the last section, the beneficial type of performance indicator is linearly related to its partial preferable probability in positively correlative manner,

Table. 3.1 Performance indicators of candidate materials of the cryogenic tank design

Material performance indicator	1	2	3	4	5	6	7	8
Aluminum 2014-T6	**75.5**	**420**	**74.2**	*2.8*	*21.4*	*0.37*	*0.16*	*1*
Aluminum 5052-O	**95**	**91**	**70**	*2.68*	*22.1*	*0.33*	*0.16*	*1.05*
Stainless steel 301 full hard	**770**	**1365**	**189**	*7.9*	*16.9*	*0.04*	*0.08*	*1.4*
Stainless steel 301 3/4 hard	**187**	**1120**	**210**	*7.9*	*14.4*	*0.03*	*0.08*	*1.5*
Ti–6Al–4V	**179**	**875**	**112**	*4.43*	*9.4*	*0.016*	*0.09*	*6.3*
Inconel 718	**239**	**1190**	**217**	*8.51*	*11.5*	*0.31*	*0.07*	*5.0*
70Cu–30Zn	**273**	**200**	**112**	*8.53*	*19.9*	*0.29*	*0.06*	*2.1*

Notice (1) toughness index; (2) yield strength, MPa; (3) Young's modulus, GPa; (4) specific gravity; (5) thermal expansion; (6) thermal conductivity; (7) specific heat; (8) relative cost

while the unbeneficial type of performance indicator contributes is linearly related to the partial preferable probability in negatively correlative manner.

Table 3.2 presents the results of partial preferable probability contributed by each material performance indicators for the seven candidate materials given in Table 3.1. The assessments for partial preferable probabilities of beneficial type or unbeneficial type of performance indicators are conducted according to Eqs. (3.1) and (3.2), or Eqs. (3.3) and (3.4), individually.

The results for the total preferable probabilities P_i of all candidate materials are given in Table 3.2 as well.

The relative value of total preferable probabilities of each candidate material in the group decides the final competitive result of the material selection comprehensively. It can be seen from the last column of Table 3.2 that "*stainless steel 301 full hard*" gets the maximum value of the total preferable probabilities P_i, so the optimal selection for the material of cryogenic tank design is the stainless steel 301 full hard, which coincides with the complex semiquantitative method by chance [1].

2. Material Selection for Automotive Brake Disk

Another example is the material selection for brake disk design [16]. Materials used for brake system should exhibit performances of stable and reliable frictional and wear properties in conditions of varying load, velocity, temperature, environmental factors, and high durability [16]. Table 3.3 presents the five candidate materials with six material performance indicators for the brake disk design [16]. Label *A* in Table 3.3 indicates compressive strength; label *B* shows friction coefficient; label *C* expresses wear rate; label *D* is specific heat, *Cp*; label *E* indicates specific gravity; and label *F* reflects relative cost. Above material performance indicators could be taken as the utility indexes in this selection.

Table. 3.2 Partial and total preferable probabilities of the candidate materials for the cryogenic tank design

Material	Partial preferable probabilities of the candidate materials							
	1	2	3	4	5	6	7	8
Aluminum 2014-T6	0.0415	0.0798	0.0754	0.2354	0.0963	0.0122	0.0714	0.1924
Aluminum 5052-O	0.0522	0.0173	0.0711	0.2388	0.0896	0.0426	0.0714	0.1908
Stainless steel 301 full hard	0.4234	0.2595	0.1920	0.0927	0.1392	0.2630	0.1667	0.1802
Stainless steel 301 3/4 hard	0.1028	0.2129	0.2134	0.0927	0.1630	0.2705	0.1667	0.1771
Ti–6Al–4V	0.0984	0.1664	0.1138	0.1898	0.2107	0.2812	0.1548	0.0305
Inconel 718	0.1314	0.2262	0.2205	0.0756	0.1907	0.0578	0.1786	0.0702
70Cu–30Zn	0.1501	0.0380	0.1138	0.0750	0.1106	0.0730	0.1905	0.1588

Material	Total	
	$P_i \times 10^8$	Rank
Aluminum 2014-T6	0.0946	6
Aluminum 5052-O	0.0798	7
Stainless steel 301 full hard	214.7950	1
Stainless steel 301 3/4 hard	56.3392	2
Ti–6Al–4 V	9.8980	3
Inconel 718	6.8412	4
70Cu–30Zn	1.1888	5

Table. 3.3 Performance indicators of the candidate materials for brake disk design

Material	*A* (MPa)	*B*	*C*	$D\ (\mathrm{J}/(\mathrm{kg}\cdot^{\circ}\mathrm{C})$	*E* ($\mathrm{ton/m^3}$)	*F*
GCI	**1293**	**0.41**	*2.36*	**0.46**	*7.2*	*1*
TiAlV	**1070**	**0.34**	*246.3*	**0.58**	*4.42*	*20*
TMC	**1300**	**0.31**	*8.19*	**0.51**	*4.68*	*20.5*
AMC 1	**406**	**0.35**	*3.25*	**0.98**	*2.7*	*2.7*
AMC 2	**761**	**0.44**	*2.91*	**0.92**	*2.8*	*2.6*

The material performance indicators *A, B,* and *D* are beneficial type of indicators, and the material performance indicators *C, E,* and *F* are unbeneficial type of material performance indicators for the material selection, respectively.

Table 3.4 presents the results of partial preferable probabilities contributed by each material performance indicators to the five candidate materials, and the results for total preferable probability of a candidate material are shown in Table 3.4 as well.

It can be seen from Table 3.4 that AMC 2 wins the competition, which coincides with the result of semiquantitative method luckily [16].

Table. 3.4 Partial and total preferable probabilities of the candidate materials for brake disk design

Material	Partial preferable probabilities of the candidate materials						Total	
	A	*B*	*C*	*D*	*E*	*F*	$P_i \times 10^5$	Rank
GCI	0.2677	0.2216	0.2513	0.1333	0.0975	0.3377	6.5428	3
TiAlV	0.2215	0.1838	0.0024	0.1681	0.1978	0.0247	0.0081	5
TMC	0.2692	0.1678	0.2453	0.1478	0.1884	0.0165	0.5077	4
AMC 1	0.0841	0.1892	0.2503	0.2841	0.2599	0.3097	9.1042	2
AMC 2	0.1576	0.2378	0.2507	0.2667	0.2563	0.3114	19.9930	1

3. Material Selection for a Cylindrical Shaft

Maleque et al. once raised the issue of material selection for cylindrical shafts [16], and there are five materials as candidates for the shaft design. Table 3.5 lists the material performance indicators of the candidate materials in the cylindrical shaft design. The candidate materials given in Table 3.5 constitute a candidate material group for selection. The material shear strength τ_f and its relative cost index C can be taken as the utility indexes in this selection. As a cylindrical shaft, it needs to have higher shear strength τ_f and lower cost index C. Therefore, the partial preferable probabilities of beneficial or unbeneficial type of indexes can be evaluated according to Eq. (3.1) or (3.3), respectively. Table 3.6 shows the evaluation results of the partial preferable probability P_{ij} and the total preferable probability P_i for each material performance indicator of the five candidate materials.

In Table 3.6, the penultimate column is the total preferable probability. The results clearly show that AISI 4340 steel has achieved the maximum value of total preferable probability in this selection. Therefore, the best choice for the cylindrical shaft design material is AISI 4340 steel, which is consistent with the result of the semiquantitative method as well.

4. Material Selection of Exhaust Manifold

Rajnish et al. studied the material selection of exhaust manifold from seven candidate materials and six material performance indicators, which are given in Table 3.7 [17]. The material performance indicators involved in Table 3.7 are taken as the utility indexes in this selection.

Table. 3.5 Performance indicators for five engineering materials of cylindrical shaft

Material	Material performance indicators	
	Strength τ_f (MPa)	Relative cost index C
Carbon fiber-reinforced composite	1140	*80*
Glass fiber-reinforced composite	1060	*40*
Al alloy (2024-T6)	300	*15*
Ti–6Al–4 V	525	*110*
AISI 4340 steel	780	*5*

Table. 3.6 Partial preferable probability P_{ij} and the total preferable probability P_i for each material of the cylindrical shaft

Material	Preferable probability of candidate material			
	P_{ij} for τ_f	P_{ij} for C	$P_i \times 10$	Rank
Carbon fiber-reinforced composite	0.2996	0.1078	0.3227	3
Glass fiber-reinforced composite	0.2786	0.2308	0.6429	2
Al alloy (2024-T6)	0.0788	0.3077	0.2426	4
Ti–6Al–4 V	0.1380	0.0154	0.0212	5
AISI 4340 steel	0.2050	0.3385	0.6938	1

Table. 3.7 Material performance indicators for material selection of exhaust manifold

Material	Performance indicators					
	SH/Bhn	CH/Bhn	SFL N/mm^2	BFL N/mm^2	UTS N/mm^2	C USC/lb
Ductile iron	220	220	460	360	880	0.342
Cast iron	200	200	330	100	380	0.171
Cast alloy steel	270	270	630	435	590	0.119
Hardened alloy steel	270	270	670	540	1190	1.283
Surface hardened alloy steel	585	240	1160	680	1580	3.128
Carburized steels	700	315	1500	920	2300	2.315
Nitrided steels	750	315	1250	760	1250	4.732

Notice SH surface hardness, *CH* core hardness, *SFL* surface fatigue limit, *BFL* bending fatigue limit, *UTS* ultimate tensile strength, *C* cost

In Table 3.7, the relative cost of the materials was the unbeneficial type of performance indicator, while other material performance indicators are beneficial type of indicators. Table 3.8 shows the evaluation results of the partial preferable probability P_{ij} and the total preferable probability P_i for each material performance indicator of the seven candidate materials.

The ranking of the evaluation results by using the new quantitative method is given in the last column of Table. 3.8.

The assessment results for the ranking of the alternatives in Table 3.8 show that carburised steel is the optimal material, and the surface hardened alloy steel is the number 2 in the ranking and cast iron is worst material among alternatives.

Table. 3.8 Partial preferable probability P_{ij} and the total preferable probability P_i for each material of the exhaust manifold

Material	Partial favorable probability of candidate material						Total	
	SH	CH	SFL	BFL	UTS	C	$P_i \times 10^6$	Rank
Ductile iron	0.0735	0.1202	0.0767	0.0947	0.1077	0.2062	1.4264	6
Cast iron	0.0668	0.1093	0.0550	0.0264	0.0465	0.2140	0.1053	7
Cast alloy steel	0.0902	0.1475	0.1050	0.1146	0.0722	0.2164	2.5017	4
Hardened alloy steel	0.0902	0.1475	0.1117	0.1423	0.1457	0.1632	5.0228	3
Surface hardened alloy steel	0.1953	0.1311	0.1933	0.1792	0.1934	0.0788	13.5224	2
Carburized steels	0.2337	0.1721	0.2500	0.2424	0.2815	0.1160	79.6054	1
Nitrided steels	0.2504	0.1721	0.2083	0.2003	0.1530	0.0054	1.4974	5

3.6 Other Applications in More Broader and General Issues

As a probability-based multi-objective optimization (PMOO) approach, its application should be widely in many fields instead of only material selections. Here in this section, the applications of PMOO in scheme selection of energy engineering, mechanical design, talent selection, purchase instrument, medical treatment, etc., are given as examples.

1. Scheme selection of energy engineering

The PMOO was used to deal with some problems in scheme selection of energy engineering including nuclear power plants (NPP) in considering the surrounding factors, the dispatching decision of reservoir flood control (RFC), and the comprehensive evaluation of design schemes for long-distance natural gas pipelines (LDNGP) [18]. Here the application of PMOO in site selection of nuclear power plants (NPP) in consideration of external human events is simply described [18].

China issued the guidance of specific document of "potential external events for the site selection of HAF0105 nuclear power plants", accordingly the on-site exploration was performed for the three plant sites; the external surrounding events related to the site selection of NPP were quantified by nine multiple indicator values, as given in Table 3.9 [19]. The schemes of the three plant sites are signed by A_1, A_2, and A_3.

There is exact meaning for each performance index in Table 3.9, which is as follows [19]:

ma—the actual distance from the airport to the nuclear power plant in unit of km;
mb—the actual distance from the explosion source to the nuclear power plant in unit of km;

Table. 3.9 Schemes of the three plant sites

Scheme	Survey value								
	ma (km)	mb (km)	mc (km)	md (km)	V_1 (m^3)	V_2 (m^3)	P_1 (%)′ 10^3	P_2 (%)′ 10^3	M (kg)
A_1	8	15	20	7	500	90	3	2.4	58
A_2	7	9	8.5	16	100	200	0.9	3.5	45
A_3	11	12	6	20	350	30	1.2	0.9	20

mc—the actual distance from the dangerous liquid source to the nuclear power plant in unit of km;
md—the actual distance from the source of the dangerous gas cloud to the nuclear power plant in unit of km;
V_1 (m^3)—the amount of storage in the facility;
P_1—the probability of any leakage or container rupture in the evaluated facility warehouse;
V_2 (m^3)—the maximum amount which can be released in the evaluated facility;
P_2—the probability of the release amount;
M—the mass of TNT or TNT equivalent of the explosive source in unit of kg.

Subsequently, the transformation of performance utility value from the survey value in Table 3.9 was performed in [19], which is cited and given in Table 3.10. The objectives C and D are the beneficial type of indexes, and the objectives Q, W, $C_b(m)$, and D_{an} are the unbeneficial type of indexes.

Table 3.11 displays the partial and total preferable probabilities for each scheme and performance utility indicator.

Table 3.11 indicates that scheme A_3 is with the maximum preferable probability in the ranking, so scheme A_3 could be chosen as the optimal decision, which means that the external surrounding factors of scheme A_3 are the more proper in the comprehensive evaluation.

2. A round log to intercept a rectangular cross-section beam

It needs to intercept a rectangular section beam from a log, how to choose the aspect ratio of the height and width of the section to make both strength and rigidity of the beam as greater as possible?

Table. 3.10 Utility value of the survey value in Table 3.9 for NPP

Scheme	Objective (utility)					
	C	D	Q	W	C_b (m)	D_{an}
A_1	0.8	1	0.7	1	14,930	0.1522
A_2	0.7	0.9	1	0.85	8936	0.0079
A_3	1	1	1	0.6	11,951	0.0045

Table. 3.11 Partial and total favorable probabilities for each performance utility and scheme

Probability	Scheme		
	A_1	A_2	A_3
P_C	0.32	0.28	0.4
P_D	0.3448	0.3104	0.3448
P_Q	0.4167	0.2917	0.2916
P_W	0.2553	0.3192	0.4255
P_{C_b}	0.2497	0.4173	0.3330
$P_{D_{an}}$	0.1049	0.4435	0.4516
$P_i \times 10^3$	0.3076	1.4969	2.5742
Rank	3	2	1

Solution: Suppose the radius of the log is r, and the angle between the connection line from the center O to the inscribed rectangular corner A is α, see Fig. 3.3, then the width b and height h of the rectangular section are

$$h = 2r \sin \alpha, \quad b = 2r \cos \alpha. \tag{3.6}$$

According to the strength conditions of the beam, under the same cross-sectional area, the larger the anti-bending section coefficient W_z of the beam the better, which is with $W_z = bh^2/6 = 4r^3 \cos \alpha \cdot \sin^2 \alpha/3$; while, according to the stiffness condition of the beam, when the cross-sectional area is the same, the larger the beam's section moment of inertia J_z the better, which is with $J_z = bh^3/12 = 4r^4 \cos \alpha \cdot \sin^3 \alpha/3$.

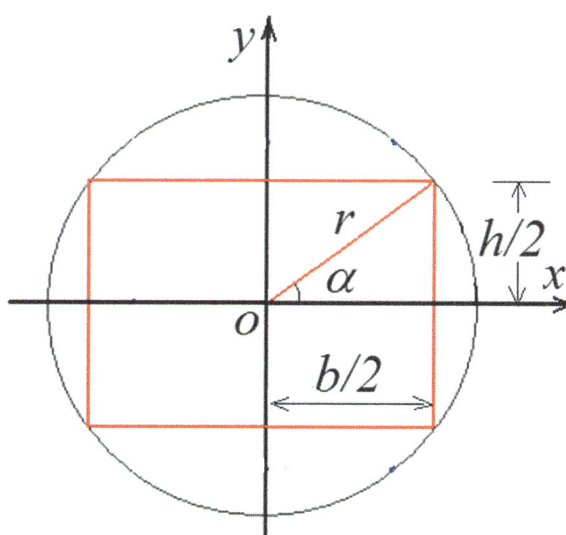

Fig. 3.3 Log rectangular beam

In this question, there involves simultaneous optimization of two responses with the characteristic of the bigger the better.

If the anti-bending section coefficient W_z is optimized individually, it leads to an aspect ratio *h*/*b* of $2^{0.5}$, which is 1.414. If the section moment of inertia J_z is optimized individually, it leads to an aspect ratio *h*/*b* of $3^{0.5}$, which is 1.732. The above two separate optimal values are different.

However, in our condition the integral optimization is conducted for both anti-bending section coefficient W_z and moment of inertia J_z simultaneously, the partial of the preferable probabilities of both anti-bending section coefficient W_z and moment of inertia J_z can be assessed according to Eq. (3.1), in addition, for continuous functions, the "summation" in the partial preferable probability evaluation process for each utility of material performance of Eq. (3.2) now becomes an integral for calculation. As a result, the optimal aspect ratio of *h*/*b* is $2.5^{0.5} = 1.581$ for the overall consideration.

3. Typical optimization of a rectangular beam with maximum strength and minimum mass

It needs to intercept a rectangular section beam from a log, how to choose the aspect ratio of the height and width of the section to make maximum strength and minimum mass?

Solution: Suppose again the radius of the log is r, and the angle between the connection line from the center O to the inscribed rectangular corner A is α, see Fig. 3.3, then the width b and height h of the rectangular section are expressed by Eq. (3.6).

According to the strength conditions of the beam, under the same cross-sectional area, the larger the anti-bending section coefficient W_z of the beam the better, which is with $W_z = bh^2/6 = 4r^3 \cos\alpha \cdot \sin^2\alpha/3$; while, according to the mass condition of the beam, when the cross-sectional area is smaller, the mass of beam will be smaller, the cross-sectional area of the beam is $A = hb = 4r^2 \cos\alpha \cdot \sin\alpha = 2r^2 \sin(2\alpha)$. Finally, in this question, it becomes to optimize the beam with the bigger W_z and smaller A. The total preferable probability $P_i(\alpha)$ of this problem is proportional to function $f(\alpha) = (1 - \sin(2\alpha)) \cdot \sin^2\alpha \cdot \cos\alpha$, *i.e.*,

$$P_i \propto f(\alpha) = (1 - \sin(2\alpha)) \cdot \sin^2\alpha \cdot \cos\alpha. \tag{3.7}$$

Consequently, the problem becomes to gain maximum of the total preferable probability P_i.

Figure 3.4 shows the variations of function $f(\alpha)$ versus α in range of $[0, \pi/2]$.

From Fig. 3.4, it can be seen that $f(\alpha)$ (also P_i) gains maximum at $\alpha = 0.42\ \pi = 1.3195$. The corresponding optimal ratio of $h/b = \text{tg}(0.42\pi) = 3.8947$ for the overall consideration.

4. Selection of talents

A unit intends to select one leader from three persons to take up the leadership role. Six attributes are used to measure, which includes health status, business status,

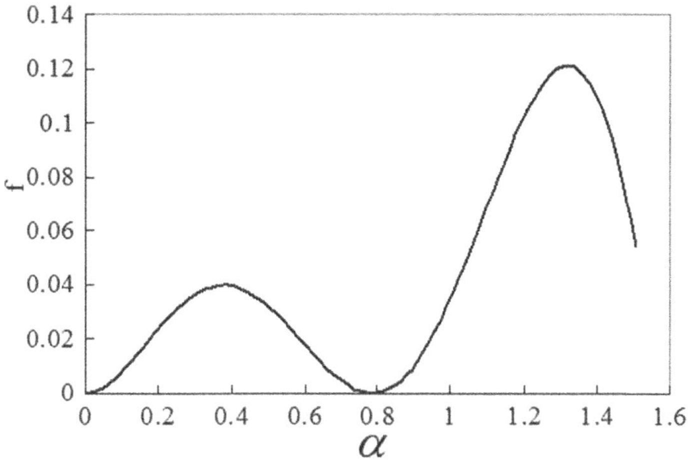

Fig. 3.4 Variations of function $f(\alpha)$ versus α in range of $[0, \pi/2]$

writing level, eloquence, policy level, and work style. They are represented by B_1, B_2, B_3, B_4, B_5, and B_6, as given in Table 3.12.

All attribute indicators have the characteristic of the bigger the better; and all of these attributes belong to the beneficial type of attribute. The evaluation results are given in Table 3.13, and the second candidate is selected.

5. Purchase instrument

It is planned to purchase a test instrument, and there are four products to be chosen. The satisfaction of each product is measured by *four* attributes, namely reliability, cost, appearance, and weight. The attribute value corresponding to each object can be quantified. The attribute values corresponding to each alternative are given in

Table. 3.12 Basic situation of the three candidates

Candidate	Health B_1	Business B_2	Writing B_3	Eloquence B_4	Policy B_5	Style B_6
No. 1	1	1	1	1	1	1
No. 2	4	4	1/3	3	1	1/7
No. 3	2	5	5	1/5	1/7	1/9

Table. 3.13 Evaluation results of selected stems

Candidate	Partial preferable probability of candidate material						Total	
	P_1	P_2	P_3	P_4	P_5	P_6	$P_i \times 10^3$	Rank
No. 1	0.1429	0.1	0.1579	0.2381	0.4666	0.7974	0.1998	2
No. 2	0.5714	0.4	0.0526	0.7143	0.4666	0.1140	0.4572	1
No. 3	0.2857	0.5	0.7895	0.0476	0.0667	0.0886	0.0317	3

Table. 3.14 Decision matrix for instrument purchase

Scheme	Instrument performance			
	Reliability F_1	Relative cost F_2	Appearance F_3	Weight F_4
X_1	7	8	9	6
X_2	6	7	8	3
X_3	5	6	7	5
X_4	4	10	6	7

Table. 3.15 Decision results of instrument purchase

Scheme	Partial preferable probability				Total	
	P_1	P_2	P_3	P_4	$P_i \times 10^3$	Rank
X_1	0.3182	0.2424	0.3000	0.2105	4.8717	2
X_2	0.2727	0.2727	0.2667	0.3684	7.3075	1
X_3	0.2273	0.3030	0.2333	0.2632	4.2289	3
X_4	0.1818	0.1818	0.2000	0.1579	1.0439	4

Table 3.14, which are represented by X_1, X_2, X_3, and X_4, respectively. Among these four objects, the reliability and appearance belong to the beneficial type of indicators, while the cost and weight belong to the unbeneficial type of indicators. The evaluation results are given in Table 3.15, and finally the second alternative instrument is selected. The selection order of the four products is $X_2 > X_1 > X_3 > X_4$.

6. Assessment of medical treatment common chemotherapy regimens for advanced non-small cell lung adenocarcinoma

The application of chemotherapy regimen for advanced non-small cell lung adenocarcinoma was assessed by Chen et al. [20]. However, the assessment was on basis of the "additive" algorithm multi-attribute utility theory. The inherent problems of personal and subjective factors in "additive" algorithm were argued previously.

Table 3.16 cited and showed the analysis consequences of utilities in the common chemotherapy regimens for advanced non-small cell lung adenocarcinoma from [20]. The efficiency is attributed to the beneficial type of index, while the cost of treatment, hospitalization days, and number of adverse reactions all belong to unbeneficial type of indexes. Table 3.17 showed the assessed results of utilities in the common chemotherapy regimens for advanced non-small cell lung adenocarcinoma by using the PMOO approach. From Table 3.17, it can be seen that Pemetrexed is competitively the proper treatment for the commonly used chemotherapy regimens for advanced non-small cell lung adenocarcinoma.

7. Comprehensive evaluation of effectiveness of fighter plane

The example is the comprehensive evaluation of effectiveness of three types of fighter plane [21], given in Table 3.18 are the relevant parameters of these fighter planes.

Table. 3.16 Analysis results of utilities in the common chemotherapy regimens for advanced non-small cell lung adenocarcinoma

Group	Efficient (%)	Treatment cost (yuan)	Hospitalization days (days)	Number of adverse reactions (person)
Gemcitabine, 14 persons	21.43	8964.26	13.91	3
Docetaxel, 22 persons	22.73	9313.33	15.33	5
Pemetrexed, 15 persons	20.00	11,893.41	5.86	1

Table. 3.17 Assessed results of utility in the common chemotherapy regimens for advanced non-small cell lung adenocarcinoma

Group	Preferable probability				Total	
	Efficient	Treatment cost	Hospitalization days	Adverse reactions	$P_i \times 10^2$	Rank
Gemcitabine, 14 persons	0.3340	0.3671	0.2557	0.3333	1.0450	2
Docetaxel, 22 persons	0.3543	0.3563	0.2058	0.1111	0.2887	3
Pemetrexed, 15 persons	0.3117	0.2767	0.5385	0.5556	2.5798	1

Table. 3.18 Relevant parameters of the three types of fighter plane

Type	Weight (ton)	Bomb load (ton)	Operational radius (km)	Cost (M ¥ RMB)
A	20.07	6.5	800	80
B	16.10	7.8	700	70
C	18.82	7.0	750	80

Table. 3.19 Utilities of the relevant parameters of the three types of fighter plane

Type	Weight U_W	Bomb W_l	Operational radius U_O	Cost U_C
A	0.80	0.70	0.80	0.7
B	0.93	0.85	0.65	0.8
C	0.85	0.80	0.71	0.7

The utilities of the relevant parameters were assessed by experts, which are given in Table 3.19. According to the meanings of the corresponding utilities [21], it has the characteristic of "the higher the better", so they belong to beneficial type of performance indexes. Table 3.20 presents the assessed consequences of the preferable probabilities together with the rank of this problem.

Table. 3.20 Assessed consequences of the preferable probabilities and rank of the three types of fighter plane

Type	P_{UW}	P_{WI}	P_{UO}	P_{UC}	$P_t \times 10^2$	Rank
A	0.3101	0.2979	0.3704	0.3182	1.0885	3
B	0.3605	0.3617	0.3009	0.3636	1.4267	1
C	0.3295	0.3404	0.3287	0.3182	1.1730	2

The result of ranking in Table 3.20 indicates that the fighter plane *Type B* is with the highest total preferable probability, which represents the highest effectiveness in the comprehensive evaluation.

3.7 Concluding Remarks

Through above description, the fundamental principle and procedure of the new probability-based multi-objective optimization in material selection is presented, which is based on the probability analysis for all possible utility indexes of material performance indicators comprehensively in viewpoint of system theory. All the utility indexes of material performance indicators are divided into beneficial type and unbeneficial (cost) type; each utility indexes of material performance indicators contributes to its partial preferable probability in positively correlative or negatively correlative manner, individually. The overall/total preferable probability of a candidate material is the product of its all possible partial preferable probabilities in the spirit of probability theory. The overall/total preferable probability of a candidate material uniquely decides the final result of the material selection comprehensively. The overall/total preferable probability of a candidate material transfers the problem of "simultaneous optimization of multiple indexes" into an overall (integrated) "optimization of single index" one.

From above study, it can be seen that the probability-based multi-objective optimization in material selection treats both beneficial type of utility indexes and unbeneficial type of utility indexes of material performance indicators equivalently and conformably, which is impersonal and without any subjective or personal factors.

The evaluation results for material selection for the cryogenic tank design, automotive brake disk, cylindrical shaft and exhaust manifold, as well as other more broader and general issues are all acceptable and conform to known, which indicates the reasonability of the new method.

References

1. M.F. Ashby, *Materials Selection in Mechanical Design*, 4th edn. (Butterworth–Heinemann, Burlington, 2011)
2. M.M. Farag, *Materials and Process Selection for Engineering Design*, 4th edn. (CRC Press, New York, 2021)
3. S. Opricovic, G.H. Tzeng, Compromise solution by MCDM methods: a comparative analysis of VIKOR and TOPSIS. Eur. J. Oper. Res. **156**, 445–455 (2004)
4. A. Shanian, O. Savadogo, Multiple-criteria decision support analysis for material selection of metallic dipolar plate for polymer electrolyte fuel cell. J. Power Source **159**, 1095–1104 (2006)
5. M.B. Babanli, F. Prima, P. Vermaut, L.D. Demchenko, A.N. Titenko, S.S. Huseynov, R.J. Hajiyev, V.M. Huseynov, in *Advances in Intelligent Systems and Computing 896, 13th International Conference on Theory and Application of Fuzzy Systems and Soft Computing—ICAFS-2018*, ed. by R.A. Aliev, J. Kacprzyk, W. Pedrycz, M. Jamshidi, F.M. Sadikoglu. Material Selection Methods: A Review (Springer Nature, Cham, 2019), pp. 929–936. https://doi.org/10.1007/978-3-030-04164-9_123
6. B.M. Ayyub, R.H. McCuen, *Probability, Statistics, and Reliability for Engineers and Scientists*, 3rd edn. (CRC Press, Taylor & Francis Group, A Chapman & Hall Book, Boca Raton, 2011) (978-1-4398-9533-7) (eBook—PDF)
7. W. Yang, S. Chon, C. Choe, J. Yang, Materials selection method using TOPSIS with some popular normalization methods. Eng. Res. Express **3**, 015020 (2021)
8. V. Modanloo, A. Doniavi, R. Hasanzadeh, Application of multi criteria decision making methods to select sheet hydroforming process parameters. Decis. Sci. Lett. **5**(3), 349–360 (2016)
9. M. Moradian, V. Modanloo, S. Aghaiee, Comparative analysis of multi criteria decision making techniques for material selection of brake booster valve body. J. Traffic. Trans. Eng. **6**, 526–534 (2019). https://doi.org/10.1016/j.jtte.2018.02.001
10. V. Modanloo, V. Alimirzaloo, M. Elyasi, Multi-objective optimization of the stamping of titanium bipolar plates for fuel cell. Int. J. Adv. Des. Manuf. Technol. **12**(4), 1–8 (2019)
11. I.Y. Kim, O. de Weck, Adaptive weighted sum method for multiobjective optimization: a new method for Pareto front generation. Struct. Multidisc. Opt. **31**(2), 105–116 (2006)
12. J. Ye, *System science* (Sichuan Academy of Social Sciences Press, Chengdu, 1987)
13. D. Wang, *Probability Theory and Mathematical Statistics* (Beijing Institute of Technology Press, Beijing, 2020)
14. G. Derringer, R. Suich, Simultaneous optimization of several response variables. J. Qual. Technol. **12**, 214–219 (1980). https://doi.org/10.1080/00224065.1980.11980968
15. L.R. Jorge, B.L. Yolanda, T. Diego, P.L. Mitzy, R.B. Ivan, Optimization of multiple response variables using the desirability function and a Bayesian predictive distribution. Res. Comput. Sci. **13**, 85–95 (2017)
16. M.A. Maleque, M.S. Salit, *Materials Selection and Design* (Springer, Heidelberg, 2013), pp.81–98
17. K. Rajnish, J. Jagadish, R. Amitava, Selection of material for optimal design using multi-criteria decision making. Proc. Mater. Sci. **6**, 590–596 (2014)
18. M. Zheng, Y. Wang, H. Teng, in *7th Virtual International Conference on Science, Technology and Management in Energy Proceedings*. Applications of "Intersection" Multi-objective Optimization in Scheme Selection of Energy Engineering (Serbia, Belgrade, 2021), pp. 89–95
19. X. Wang, S. Zou, B. Pang, The assessing method on site—choosing of NPP about outside artificial event based on fuzzy optimal selection. Value Eng. **4**, 8–10 (2009). https://doi.org/10.14018/j.cnki.cn13-1085/n.2009.04.002
20. M. Chen, X. Lu, Q. Zhu, L. Xu, Evaluation of common chemotherapy regimens in advanced non-small cell lung adenocarcinoma based on multi-attribute utility theory. Chin. J. Drug Appl. Monitor. **18**(1), 1–4 (2021)
21. L. Yang, X. Gao, J. He, A comprehensive method for effectiveness evaluation of a fighter plane. J. Northwestern Polytech. Univ. **21**(1), 42–45 (2003)

Chapter 4
Robustness Evaluation with Probability-Based Multi-objective Optimization

Abstract Under condition of utility with uncertainty or interval number, the assessment of each utility involves two independent responses, i.e., the mean value and the variance of utility of the performance index. In the assessment, it adopts the separate models for the mean value and the variance of utility of the performance index in the optimization, and thus the product of both parts of partial preferable probabilities of the mean (or central) value and the mean variance of utility of the performance index is the actual entire partial preferable probability of the corresponding utility.

4.1 Introduction

In the last chapter, the probability-based multi-objective optimization (PMOO) method was developed in an attempt to reflect the intrinsic essence of "simultaneous optimization of multi-performance utility indexes" in viewpoint of system theory. The new concept of preferable probability was proposed to represent the preferable degree of performance utility index of a candidate in the optimization. In PMOO, all performance utility indicators of candidates are attributed into two types preliminarily, i.e., the beneficial type or unbeneficial type according to their functions in the optimization; each performance utility indicator of the candidate contributes one partial preferable probability quantitatively, and furthermore, the product of all partial preferable probabilities forms the overall/total preferable probability of the candidate in the spirit of probability theory, which is the uniquely decisive index in the selection process. Thus, the overall/total preferable probability transfers the multi-objective optimization problem into an overall (integrated) single-objective one [1, 2].

Historically, a British statistician and geneticist named Ronald Aylmer Fisher developed the use of statistics in genetics and biomathematics. In 1919, Fisher was offered two posts simultaneously. He accepted the post at Rothamsted where he made many contributions both to statistics, in particular the design and analysis of experiments, and to genetics. Rothamsted Agricultural Experiment Station was the oldest agricultural research institute in the UK, established in 1837 to study the effects of nutrition and soil types on plant fertility.

M. Zheng et al., *Probability-Based Multi-objective Optimization for Material Selection*, https://doi.org/10.1007/978-981-99-3939-8_4

At Rothamsted, he studied the design of experiments by introducing the concept of randomization and the analysis of variance, as well as procedures now used throughout the world. Fisher's main idea was to arrange an experiment as a set of partitioned sub-experiments that differ from each other in having one or several factors or treatments applied to them. The sub-experiments were designed in such a way as to permit differences in their outcome to be attributed to the different factors or combinations of factors by means of statistical analysis. This was a notable advance as compared to the existing approach of varying sole one factor at a time in an experiment, which was a relatively insufficient procedure.

The concept of likelihood was introduced by him in 1921. The likelihood of a parameter indicates proportional to the probability of the data and gives a function that has a single maximum value usually; he called it the maximum likelihood.

Moreover, he gave a new definition of statistics in 1922. Its purpose was the reduction of data with three fundamental problems, i.e.,

i. specification of the kind of population that the data came from;
ii. estimation; and
iii. distribution.

Fisher published a number of important books; the book *Statistical Methods for Research Workers* (1925) ran to many editions in particular. It became a handbook for the methods for the design and analysis of experiments. The contributions of Fisher included the development of methods suitable for small samples and the discovery of the precise distributions of many sample statistics. Later, Fisher published *The design of experiments* (1935) and *Statistical tables* (1947), all these works laid the foundation for modern statistics [3].

In industrial productions, quality improvement of products and optimization of processes are persistently needed by both sides of customers and producers. In 1950s, Taguchi once initiated a subject and structure to the design and assessment of experiments to improve quality of products by means of optimal design with efficient consumption [4]. According to Taguchi's procedure, noise factors are incorporated in the assessment of experiment layout, which attempts to get an insensitive status of products and processes with less influence of uncontrollable (noise) factors. His orthogonal experiment design is used to study the effects of noise factors with smaller number of experiments, which leads to a preferable performance with the mean value close to the target and a reduced variation around the mean [4]. The main point is braw, which aims to let the output response close to the prechosen target with less variability. The controllable factors are called control variables. It is assumed that the existence of uncontrollable factors, i.e., noise factors or noise variables, leads to the majority of variability around the target. While in the product design or process operation, noise factors are uncontrollable and inevitable [5].

Consequently, the term robust parameter design entails designing the system in order to gain robustness (insensitivity) to the inevitable changes of the noise variables. Furthermore, Taguchi adopted a factor called "signal-to-noise ratio" (SNR) to depict robustness. Three kinds of specific goals suggested by Taguchi are commonly used, which include (1) the target is the best; (2) the larger the better; and (3) the smaller the better.

As to cases in which the standard deviation of response is linearly related to the mean, SNR for "the target is the best" condition suggested by Taguchi is given by

$$\mathrm{SNR} = -10 \cdot \log\left(\overline{y}^2/s^2\right) \quad (4.1)$$

In Eq. (4.1), it takes maximum of SNR robust optimization; $\overline{y}$ is the arithmetic mean value of the test data, and s stands for the mean deviation, commonly standard error.

In general, the mean value of the test data $\overline{y}$ and the mean deviation s are independent responses for a set of actual experiments or processes in principle.

However, the expression of SNR in Eq. (4.1) simply casts the two responses into one model, and the optimization of the maximum for the unique response SNR is not equivalent to the simultaneous optimization of both minima of s and $\overline{y}$ closing to the target individually.

The worst cases are that in the conditions of "the larger the better" and "the smaller the better", the expressions of *SNRs* proposed by Taguchi even excluded the role of the mean deviation; see Eqs. (4.2) and (4.3),

$$\mathrm{SNR} = -10 \cdot \lg(1/m)(1/y_1^2 + 1/y_2^2 + \cdots + 1/y_m^2) \text{ for "the larger the better"}, \quad (4.2)$$

$$\mathrm{SNR} = -10 \cdot \lg(1/m)(y_1^2 + y_2^2 + \cdots + y_m^2) \text{ for "the smaller the better"}. \quad (4.3)$$

In Eqs. (4.2) and (4.3), m is the number of experiment times.

This was the point that is usually castigated by many statistical scientists [6–10]!

Although the starting point of Taguchi's approach is to develop an easy-to-use performance criterion to the robust parameter design by taking the mean value and variance into consideration simultaneously [6–10], the algorithm of SNR employed is not an appropriate manipulation.

Moreover, statisticians suggested using separate models to deal with both mean value and variance of response in robustness study [6–10]. Therefore, the optimization of both $\overline{y}$ closing to the target and minima of s should be conducted with individual models in proper manner at the same time. Box preferred to study robustness via Bayes's theorem [11–13].

In this chapter, the newly developed PMOO methodology is extended to contain robust optimization of data with uncertainty in material engineering, where both the mean $\overline{y}$ and the variance s of response are taken into account by using separate models. Furthermore, some examples are given, which include energy consumption in the melting process with orthogonal array design, robust optimization of four different process schemes in the machining process of the electric globe valve body, and robust optimization for cutting process of ferrite-bainite dual-phase steel.

4.2 Extension of Probability-Based Multi-objective Optimization to Contain Robustness

In PMOO, the performance utility indicators of candidates is preliminarily divided into beneficial or unbeneficial types according to their functions in the optimal process, each performance utility indicator of the candidate contributes to one partial preferable probability quantitatively, and in the spirit of the probability theory the product of all partial preferable probabilities gives the overall/total preferable probability of a candidate; the overall/total preferable probability of the candidate is its uniquely deterministic index in the selection process, which thus transfers the multi-objective optimization problem into a single-objective one [1, 2].

In considering the uncertainty and variance of experimental data, the traditional MOO cannot be simply employed directly. The performance indexes of candidates are well defined in traditional MOO without any uncertainty. However, the practical cases are not always like this; for example, when one conducts an experiment for sixteen times, one could not get his experimental data in the sixteen times keeping exactly the same value in general, while both the arithmetic mean value and the mean deviation of the sixteen data can be seen as representatives for one's experiments. Besides, in some other cases, it is quite often that the performance indexes and attributes are vague, which leads to inexact numerical data instead of well-defined value. In order to deal with such problems that includes uncertain elements, an appropriate approach is undoubtedly still needed. Here in this chapter, we develop an extension for the newly proposed PMOO to contain the variance such that the probability-based multi-objective robust optimization with dispersed data is formulated.

In general, if an element U_{ij} is with uncertainty, which can be written as

$$U_{ij} = U_{ij0} + \delta U_{ij}. \tag{4.4}$$

In Eq. (4.4), U_{ij0} expresses the arithmetic mean value of the uncertain element U_{ij}, and δU_{ij} is the deviation of the performance index U_{ij} from U_{ij0}.

The arithmetic mean value U_{ij0} represents the main function of the performance of a candidate, which contributes one part of partial preferable probability quantitatively according to its type of being either beneficial or unbeneficial, the type of this performance is related to its role or function in the selection.

(1) **For cases of "the larger the better" and "the smaller the better"**

For the beneficial type of performance, it makes its contribution of one part of partial preferable probability positively in linear manner; while the unbeneficial type of performance, it makes its contribution of one part of partial preferable probability negatively in linear manner [1, 2].

Under condition of the uncertain element U_{ij}, the beneficial type of the arithmetic mean value U_{ij0} of the uncertain element U_{ij} makes one part of the performance index according to

$$P_{ij1} = a_{j1} U_{ij0}, \quad i = 1, 2, \ldots, n; \quad j = 1, 2, \ldots, m. \tag{4.5}$$

In Eq. (4.5), P_{ij1} is the one part of the partial preferable probability of the beneficial type of utility index U_{ij0}; n represents the total number of candidates in the candidate group involved; m indicates the total number of the performance utility indexes of each candidate in the group; α_{j1} represents the normalized factor of the jth utility index of the candidate performance indicator, $\alpha_{j1} = 1/(n\overline{U_{jo}})$, $\overline{U_{j0}}$ represents the arithmetic mean value of the utility index U_{ij0} of the performance indicator in the candidate group involved,

$$\overline{U_{j0}} = \frac{1}{n} \sum_{i=1}^{n} U_{ijo}. \tag{4.6}$$

While, for the unbeneficial type of performance utility index, U_{ij0} contributes one part of its partial preferable probability of the performance according to

$$P_{ij1} = \beta_{j1}(U_{j0\max} + U_{j0\min} - U_{ij0}), \quad i = 1, 2, \ldots, n; \quad j = 1, 2, \ldots, m. \tag{4.7}$$

$U_{j0\min}$ and $U_{j0\max}$ in Eq. (4.7) indicate the minimum and maximum values of the performance utility indices U_{ij0} of the candidate performance indicator in the group, respectively; β_{j1} indicates the normalized factor of the jth utility indexes of the candidate performance indicator, $\beta_{j1} = 1/[n(U_{j0\min} + U_{j0\max}) - n\overline{U_{j0}}]$.

In general, the deviation δU_{ij} is the unbeneficial type of the performance index in assessment due to its characteristic of deviation, which may have the feature of "the lower the better" in general. The deviation δU_{ij} makes its contribution to the other part of the uncertain element U_{ij} by P_{ij2}, which follows Eq. (4.8)

$$P_{ij2} = \beta_{j2} \times (\delta U_{j\max} + \delta U_{j\min} - \delta U_{ij}), \\ i = 1, 2, \ldots, n; \quad j = 1, 2, \ldots, m. \tag{4.8}$$

$\delta U_{j\min}$ and $\delta U_{j\max}$ in Eq. (4.8) represent the minimum and maximum values of the performance utility indexes δU_{ij} of the candidate performance indicator in the group, respectively, and β_{j2} is the normalized factor of the jth utility indices of performance indicator deviation of the candidate δU_{ij}, $\beta_{j2} = 1/[n(\delta U_{j\min} + \delta U_{j\max}) - n\overline{\delta U_j}]$, and $\overline{\delta U_j} = \frac{1}{n}\sum_{i=1}^{n} \delta U_{ij}$.

(2) **For case of "the target is the best"**

As to the case of "the target is the best", let the target value be U_{jT}, then the difference between the arithmetic mean value U_{ij0} and the target value U_{jT} can be taken as the actual utility, says,

$$\Delta U_{ij} = |U_{ij0} - U_{jT}|, \tag{4.9}$$

which leads to the first part of the partial preferable probability P_{ij1} and ΔU_{ij} belongs to the unbeneficial type of performance index, while the other part δU_{ij} still belongs to unbeneficial type of performance and contributes to the second part of the partial preferable probability P_{ij2}.

(3) **Entire partial preferable probability of the uncertain element U_{ij}**

Moreover, the entire partial preferable probability of the uncertain element U_{ij} is the product of both parts logically, i.e.,

$$P_{ij} = (P_{ij1} \times P_{ij2}). \tag{4.10}$$

The entire partial preferable probability P_{ij} contains all information of the uncertain element U_{ij} comprehensively, which is the overall representative of the uncertain element U_{ij} in the selection process integrally.

Furthermore, according to probability theory, the overall/total preferable probability of the ith candidate in a multi-objective optimization problem is comprehensively the product of its partial preferable probability P_{ij} of each utility index of the candidate performance indicator in the overall consideration due to the "simultaneous optimization" of multiple objectives [1], i.e.,

$$P_i = P_{i1} \cdot P_{i2} \cdots P_{im} = \prod_{j=1}^{m} P_{ij}. \tag{4.11}$$

The overall/total preferable probability of a candidate is the uniquely decisive index in the overall selection process comprehensively, which thus transfers a multi-objective optimization problem (MOOP) into a single-objective optimization one. The remarkable characteristic of the new probability-based multi-objective optimization is that the treatment for both beneficial type of utility index and unbeneficial type of utility index is equivalent and conformable; on the other hand, it is without any artificial or subjective scaling factors involved in the process evidently.

4.3 Application of the Extended PMOO in Evaluation of Optimal Problems with Variance of Data in Material Engineering

In the following section, three examples are given, which involves energy consumption in the melting process with orthogonal array design, robust optimization of four different process schemes in the machining process of the electric globe valve body, and robust optimization for cutting process of ferrite-bainite dual-phase steel.

1. **Robust optimization for electric energy saving of a foundry process**

In foundry process, electric furnaces are widely used in general, which includes rotary furnaces, cupola furnaces, and induction furnaces. The induction furnace is frequently used to melt a massive amount of steel. The electric energy consumed for melting 1 ton of metal is within the range of 600–680 kWh/ton [14]. Deshmukh et al. employed an orthogonal array experiment in their melting process for foundry to optimize the process parameters with a "signal-to-noise ratio" effect [14]. The study was conducted by varying the process parameters, which aims to reduce electric energy consumption with robust optimization. L9 orthogonal array was employed to perform the designed experiment for controlling factors of weight percentage of bundled steel, loose steel, and uncleaned steel, as is given in Table 4.1.

Each option of the nine designed schemes was performed five times to reveal the variations that might be caused by noise factors. Table 4.2 shows the test result of electric energy consumption from these designed schemes.

In this problem, the optimization is to save electric energy consumption, so the mean value of the electric energy consumption in Table 4.2 belongs to an unbeneficial type of performance index, thus Eq. (4.7) is appropriate to conduct the assessment of its partial preferable probability. Besides, Eq. (4.8) is appropriate to conduct the assessment of the deviation contribution to the partial preferable probability. Finally, the entire partial preferable probability of each scheme is evaluated by Eq. (4.10). The results of the assessments are given in Table 4.3. The labels $P_{deviation}$ and P_{mean} in Table 4.3 indicate one part of partial preferable probability of the deviation value and the mean value of electric energy consumption, individually; P_{entire} presents the entire partial preferable probability of electric energy consumption, which uniquely decides the ranking of each scheme in Table 4.3.

Table 4.1 Designed weight percentage of charging material

Scheme	Bundled steel (% by weight)	Uncleaned steel (% by weight)	Loose steel (% by weight)
1	12.5	50	37.5
2	33	33	33
3	37.5	50	12.5
4	50	50	0
5	12.5	37.5	50
6	50	37.5	12.5
7	50	0	50
8	33	33	33
9	37.5	12.5	50

Table 4.2 Test result of electric energy consumption from the designed scheme

Scheme	Test data (kWh)					Representative data	
	1	2	3	4	5	Mean	Deviation
1	110	108	104	112	131	113.0	9.3808
2	109	121	114	111	120	115.0	4.7749
3	112	118	110	120	115	115.0	3.6878
4	98	112	104	102	106	104.4	4.6303
5	117	113	108	112	109	111.8	3.1875
6	121	107	113	116	109	113.2	4.9960
7	114	110	112	118	108	112.4	3.4409
8	116	104	109	112	110	110.2	3.9192
9	110	109	107	118	112	111.2	3.7630

Table 4.3 Assessed results of the preferable probability of all schemes and their ranking

Scheme	$\underline{P}_{\text{mean}}$	$P_{\text{deviation}}$	$P_{\text{entire}} \times 10^2$	Rank
1	0.1099	0.0447	0.4910	9
2	0.1078	0.1093	1.1778	7
3	0.1078	0.1245	1.3421	5
4	0.1188	0.1113	1.3215	6
5	0.1111	0.1315	1.4612	1
6	0.1097	0.1062	1.1641	8
7	0.1105	0.1280	1.4138	2
8	0.1128	0.1212	1.3672	4
9	0.1117	0.1234	1.3792	3

Table 4.3 indicates that scheme 5 is the optimal one, lower electric energy with less deviation is consumed in this scheme, so it is with good robustness. The optimal weight percentages of the controlling factors of bundled steel, uncleaned steel, and loose steel are 12.5%, 37.5%, and 50% in steel melting process, individually; the scheme 7 is ranked No. 2, which is close to scheme 5 with the weight percentages of the controlling factors of bundled steel, uncleaned steel, and loose steel are 50%, 0%, and 50%, respectively.

2. **Robust optimization for mechanical processing schemes with multiple objectives with interval number**

A multi-objective robust decision making for a mechanical processing scheme with the interval number was conducted by Han et al. [15]; four schemes for the machining process of an electric globe valve body are investigated competitively, which is reanalyzed here again.

Table 4.4 Technical parameters of the four schemes

Scheme	Time for product *A* (min)	Rate of qualified products *B* (%)	Total cost *C* (¥ RMB)	Material consump. *D* (¥ RMB)
1	[40, 51]	[96, 98]	[238, 285]	[82.6, 114.5]
2	[48, 59]	[91, 95]	[254, 303]	[92.4, 123.3]
3	[50, 62]	[89, 92]	[258, 310]	[94.2, 126.1]
4	[42, 56]	[92, 96]	[245, 292]	[86.8, 116.9]

Scheme	Electric energy consump. *E* (°)	Solid waste *F* (kg)	Waste liquid discharge *G* (L)
1	[18.6, 21.5]	[0.86, 0.97]	[2.8, 3.1]
2	[19.8, 23.2]	[0.95, 1.22]	[2.9, 3.5]
3	[20.3, 25.2]	[1.07, 1.28]	[3.1, 3.9]
4	[19.1, 22.3]	[0.92, 1.15]	[2.9, 3.3]

Table 4.5 Partial preferable probability and the total preferable probability of each scheme, together with their ranking

Scheme	Partial preferable probability × 10						
	A	*B*	*C*	*D*	*E*	*F*	*G*
1	0.7464	0.9419	0.6744	0.6462	0.7695	1.0923	0.9327
2	0.6396	0.4516	0.6096	0.6114	0.6472	0.3806	0.5415
3	0.5631	0.6589	0.5612	0.5784	0.4016	0.5355	0.2948
4	0.5494	0.4563	0.6579	0.6646	0.7023	0.5448	0.7812

Scheme	Total	
	$P_t \times 10^9$	Rank
1	24.0184	1
2	1.4363	3
3	0.7636	4
4	3.2758	2

The technical parameters of the four schemes are given in Table 4.4. In this problem, only the rate of the qualified product belongs to the beneficial type of the performance index, others are attributed to unbeneficial type. The partial preferable probability and the total preferable probability of each scheme together with the overall ranking are given in Table 4.5 comparatively. Table 4.5 indicates that scheme 1 is the optimal one with good robustness.

3. **Robust optimization for cutting process of ferrite-bainite dual-phase steel**

Hegde et al. once designed the cutting process of ferrite-bainite dual-phase steel (AISI1040 F-B) robustly [16]. In the experiments, the tool life and surface roughness of sample were taken as their simultaneous optimization goals to optimize the input parameters of ferrite-bainite dual-phase steel (AISI1040 F-B), such as heating temperature (A), cutting speed (B), feed speed (C), and cutting depth (D) of material heat treatment. Each input parameter has three levels, and each experimental condition uses three samples in the experiment [16]. They used Taguchi $L_9(3^4)$ for experimental design, and the arithmetic value and standard deviation of the experimental results are listed in Table 4.6. In Table 4.6, T_{LA} (E) and δT_L (F), respectively, represent the arithmetic value and standard deviation of tool life, while S_{RA} (G) and δ_{SR} (H) reflect the arithmetic value and standard deviation of sample surface roughness.

Table 4.7 gives the evaluation and ranking results of the preferable probability of cutting experiments designed by $L_9(3^4)$ for this problem. In the evaluation, according to the requirements of robust optimization, only quantity T_{LA} is beneficial type, and all other responses have the characteristics of unbeneficial type.

The evaluation results in Table 4.7 show that the No. 9 experimental scheme has the highest total preferable probability value P_i. Therefore, the configuration of robust

Table 4.6 Design and experimental results of cutting parameters of dual-phase steel with $L_9(3^4)$

No.	Input parameter			
	A (°C)	B (m/min)	C (mm/rev)	D (mm)
1	750	80	0.13	0.2
2	750	115	0.15	0.4
3	750	150	0.18	0.6
4	770	80	0.15	0.6
5	770	115	0.18	0.2
6	770	150	0.13	0.4
7	790	80	0.18	0.4
8	790	115	0.13	0.6
9	**790**	**150**	**0.15**	**0.2**

No.	Objective			
	Arithmetic value of tool life	Standard deviation of tool life	Arithmetic value of sample surface roughness	Standard deviation of sample surface roughness
	T_{LA}, E (s)	δT_L, F (s)	S_{RA}, G (μm)	δS_R, H (μm)
1	2646	29.4618	4.2633	0.0416
2	1907	1.7321	4.0833	0.1589
3	994	3.6056	2.6233	0.0551
4	1464	6.9282	4.07	0.0458
5	2168.333	16.0728	3.11	0.0854
6	1172	19	2.5567	0.0551
7	1528.333	2.0817	3.1067	0.0902
8	700	4.3589	2.42	0.0889
9	**1297.333**	**2.0817**	**2.22**	**0.02**

Table 4.7 Evaluation and ranking of preferable probability for cutting of dual-phase steel with $L_9(3^4)$

No.	Partial preferable probability				Total preferable probability	Rank
	P_E	P_F	P_G	P_H	$P_i \times 10^4$	
1	0.1907	0.0089	0.0743	0.1417	0.1778	9
2	0.1374	0.1508	0.0803	0.0206	0.3432	8
3	0.0716	0.1412	0.1291	0.1277	1.6679	3
4	0.1055	0.1242	0.0807	0.1373	1.4523	4
5	0.1563	0.0774	0.1128	0.0965	1.3162	5
6	0.0845	0.0624	0.1313	0.1277	0.8842	6
7	0.1101	0.1490	0.1129	0.0915	1.6961	2
8	0.0504	0.1373	0.1359	0.0929	0.8743	7
9	**0.0935**	**0.1490**	**0.1426**	**0.1640**	**3.2564**	**1**

design is near the parameters of experiment scheme No. 9. In addition, Table 4.8 shows the results of range analysis of the total preference probability of each group of schemes given in Table 4.7. Table 4.8 shows that the order of impact intensity of input variables is $A > B > C > D$ and reveals that the optimal configuration is $A_3B_3C_2D_1$, which is the experiment scheme No. 9. While Hegde et al. used ANOVA technology to statistically analyze the relative contributions of various factors to T_L and S_R, their optimization results were close to the experimental scheme No. 2 [16]. Obviously, from the respect of probability theory, the result of experiment scheme No.9 is superior to that of experiment scheme No. 2.

For the details of analysis of the hybrid of probability-based multi-objective optimization with orthogonal experimental design, please see Sect. 6.2.

Table 4.8 Range analysis for total preferable probability of experimental design of cutting for dual-phase steel (AISI1040 F-B) with $L_9(3^4)$

Level	Parameter			
	A	*B*	*C*	*D*
1	0.7296	1.1087	0.6454	1.5835
2	1.2176	0.8446	1.6840	0.9745
3	1.9423	1.9362	1.5600	1.3315
Range	1.2126	1.0916	1.0386	0.6090
Impact order	1	2	3	4
Optimal configuration	A_3	B_3	C_2	D_1

4.4 Conclusion

The extension of the probability-based multi-objective optimization in considering interval value is successful, which is more appropriately used to conduct a robust optimization of problem with uncertainty objectively in material engineering.

Being a very significant technology to improve quality of products and optimize processes, robust optimization design is useful for both customers and producers; the extension here will be a starting point to the relevant research and process optimization.

References

1. M. Zheng, Y. Wang, H. Teng, A new "intersection" method for multi-objective optimization in material selection. Tehnički Glas. **15**(4), 562–568 (2021). https://doi.org/10.31803/tg-20210901142449.
2. M. Zheng, H. Teng, Y. Wang, Robust design in material machining on basis of probability multi-objective optimization. Materialwiss. Werkstofftech. **54**, 180–185 (2023). https://doi.org/10.1002/mawe.202200162
3. R. A. Fisher, *The Design of Experiment* (Oliver and Boyd, Edinburgh, 1935).
4. R.K. Roy, *A Primer on the Taguchi Method*, 2nd edn. (Society of Manufacturing Engineers, Dearborn, 2010). ISBN-13: 978-0872638648
5. R.H. Myers, D.C. Montgomery, C.M. Anderson-Cook, *Response Surface Methodology: Process and Product Optimization Using Designed Experiments*, 4th edn. (Wiley, Hoboken, 2016). ISBN: 978-1-118-91601-8
6. G. Box, Signal-to-noise ratios, performance criteria, and transformations. Technometrics **30**(1), 1–17 (1988). https://doi.org/10.2307/1270311
7. G.E.P. Box, R.D. Meyer, Dispersion Effects from Fractional Designs. Technometrics **28**(1), 19–27 (1986). https://doi.org/10.1080/00401706.1986.10488094
8. W.J. Welch, T.-K. Yu, S.M. Kang, S.M.J. Sacks, Computer experiments for quality control by parameter design. J. of Qual. Techn. **22**(1), 15–22 (1990). https://doi.org/10.1080/00224065.1990.11979201
9. W.J. Welch, R.J. Buck, J. Sacks, H.P. Wynn, T.J. Mitchell, M.D. Morris, Screening, Predicting, and Computer Experiments. Technometrics **34**(1), 15–25 (1992). https://doi.org/10.1080/00401706.1992.10485229
10. V. N. Nair, B. Abraham, J. MacKay, G. Box, R. N. Kacker, T. J. Lorenzen, J. M. Lucas, R. H. Myers, G. G. Vining, J. A. Nelder, M. S. Phadke, J. Sacks, W. J. Welch, A. C. Shoemaker, K. L. Tsui, S. Taguchi, C. F. Jeff Wu, Taguchi's Parameter Design: A Panel Discussion. Technometrics **34**(2), 127–161(1992). https://doi.org/10.2307/1269231.
11. G. E. P. Box, G. C. Tiao, A further look at robustness via Bayes's theorem. Biometrika 49 (3,4), 419–432 (1962).
12. G. E. P. Box, G. C. Tiao, A Bayesian approach to the importance of assumptions applied to the comparison of variances. Biometrika **51**(1, 2), 153–167 (1964).
13. G. E. P. Box, G. C. Tiao, A Note on Criterion Robustness and Inference Robustness, Biometrika, **51**(1, 2), 169–173 (1964).
14. R. Deshmukh, R. Hiremath, Societal application of Taguchi method for optimization of process parameters in the Melting Process in the Foundry, in *Techno – Societal*, ed. by P. Pawar, B. Ronge, R. Balasubramaniam, A. Vibhute, S. Apte (Springer, Cham,2020). pp. 215–221. https://doi.org/10.1007/978-3-030-16962-622.

15. Z. Han, W. Shan, T. He, Multi-objective Robust Decision-Making of Machining Process Scheme Based on Interval Number. Modern Manufacturing Engineering **5**, 98–101 (2020). https://doi.org/10.16731/j.cnki.1671-3133.2020.05.015.
16. A. Hegde, J. Hindi, B.M. Gurumurthy, S. Sharma, A. Ki, Machinability study and optimization of tool life and surface roughness of Ferrite-Bainite dual phase steel. J. of Appli. Eng. Sci. **20**(2), 358–364 (2022). https://doi.org/10.5937/jaes0-32927

Chapter 5
Extension of Probability-Based Multi-objective Optimization in Condition of the Utility with Desirable Value

Abstract It represents the extension of probability-based multi-objective optimization in condition of the utility with desirable value, which includes the types of "one side desirability problem" and "one range desirability problem"; thereafter, the evaluations of partial and total preferable probabilities of the multi-objective optimization experiment design are conducted according to the common procedure of the probability multi-objective optimization. Finally, regression analysis is employed for the total preferable probability to get its maximum and optimal status with desirable response variable. Application examples of the experimental designs of maximizing yield with constraints of viscosity and molecular weight and the maximizing conversion rate with constraints of desirable thermal activity are given in detail; satisfied results are obtained.

5.1 Introduction

In industrial production, architecture building, chemical reaction, transportation, banking, social activities, etc., an eternal topic is optimization. Many performances or attributes are involved likely, which need to be fully taken into account in the analysis. In some conditions, an optimal status needs to meet specific demands of performance indexes or response variables, i.e., desired value, while these demands of performance indexes are even conflicting each other. The proper algorithm to deal with this issue is to consider all responses approaching to their desirable values simultaneously.

In this chapter, the probability-based multiple objective optimization is extended to include the condition of the utility with desirable value.

Besides the beneficial or unbeneficial types of performance utility indicators of candidates, the condition of desirable values for performance indexes or response variables can be considered as a third type of optimal requirement for performance indicators [1, 2], which has the feature of the desired target being the best. In 1980s, Derringer and Suich once addressed this problem by introducing a desirability function and desirable transformation with an assignable weighting exponent [3], instead of probability-based approach.

M. Zheng et al., *Probability-Based Multi-objective Optimization for Material Selection*, https://doi.org/10.1007/978-981-99-3939-8_5

Impersonally, the extension of probability-based multi-objective optimization in condition of the performance response with desirable value might be appropriate.

5.2 Assessments of Partial and Overall Preferable Probability for Performance Response with Desirable Value in the Probability-Based Multiple Objectives Optimization

5.2.1 One Range Desirable Value Problem

The utility was employed to reflect the usefulness of a performance response or product in reference to the expectations of customers or users [4]. The methodology of utility approach is to transform the performance response of each qualified characteristic into a common index.

Under condition of one range desirability, the response variable R_{ij} is commonly with a desirable response range, such as range of $[\alpha, \beta]$. In such case, the utility of a performance response value R_{ij} falling within the range of $[\alpha, \beta]$ is with the same value, and zero for those performance response value R_{ij} falling outside range of $[\alpha, \beta]$, i.e.,

$$U_{ij} = \begin{cases} 1, R_{ij} \in [\alpha, \beta]; \\ 0, R_{ij} \notin [\alpha, \beta]; \end{cases} \quad i = 1, 2, \ldots, n; \quad j = 1, 2, \ldots, m. \tag{5.1}$$

According to assessment algorithm of probability-based multi-objective optimization in Chapter 3, the partial preferable probability P_{ij} of utility U_{ij} should possess a constant value v_j within range of $[\alpha, \beta]$, and zero value outside range of $[\alpha, \beta]$, i.e.,

$$P_{ij} = \begin{cases} v_j, R_{ij} \in [\alpha, \beta]; \\ 0, R_{ij} \notin [\alpha, \beta]; \end{cases} \quad i = 1, 2, \ldots, n; \quad j = 1, 2, \ldots, m. \tag{5.2}$$

R_{ij} represents the *j–th* performance response indicator of the *i–th* candidate; P_{ij} is the partial preferable probability of the one range desirable performance response variable R_{ij}; n indicates the total number of candidates in the candidate group involved; m expresses the total number of performance indicators of each candidate in the group; v_j is value of the partial preferable probability P_{ij} of the *j–th* response variable R_{ij} within range of $[\alpha, \beta]$.

In light of probability theory [5], the summation of each P_{ij} for the index i in *j–th* performance response factor is normalized to 1, i.e., $\sum_{i=1}^{n} P_{ij} = 1$, it thus naturally results in

$$\sum_{i=1}^{n} v_j = 1, \quad v_j = 1/l, \tag{5.3}$$

l is the number of the value of the performance response variable R_{ij} falling in the range of the $[\alpha, \beta]$.

5.2.2 One Side Desirable Value Problem

Under condition of one side desirability, the response variable R_{ij} has a desirable response limit, i.e., within range of $[0, \beta]$, which is a special case of one side desirable condition as letting $\alpha = 0$. In this case, the utility of a performance response value R_{ij} falling within the range of $[0, \beta]$ is with the same value as well and zero for those performance response value R_{ij} falling outside range of $[0, \beta]$, i.e.,

$$U_{ij} = \begin{cases} 1, R_{ij} \in [0, \beta]; \\ 0, R_{ij} \notin [0, \beta]; \end{cases} \quad i = 1, 2, \ldots, n; \quad j = 1, 2, \ldots, m. \tag{5.4}$$

Correspondingly, the partial preferable probability P_{ij} of utility U_{ij} might possess a constant value w_j within range of $[0, \beta]$ and zero value outside range of $[0, \beta]$, i.e.,

$$P_{ij} = \begin{cases} w_j, R_{ij} \in [0, \beta]; \\ 0, R_{ij} \notin [0, \beta]; \end{cases} \quad i = 1, 2, \ldots, n; \quad j = 1, 2, \ldots, m. \tag{5.5}$$

Analogically, in accordance with probability theory [5], the summation of each P_{ij} for the index i in j-th performance factor is normalized 1, i.e., $\sum_{i=1}^{n} P_{ij} = 1$, it thus leads to following result,

$$\sum_{i=1}^{n} w_j = 1, \quad w_j = 1/k, \tag{5.6}$$

k is the number of the value of the performance response variable R_{ij} falling within range of the $[0, \beta]$.

Similarly, the problem of one side desirability within range of $[\alpha, \propto]$ can be treated in the same way of above procedure.

As the partial preferable probability P_{ij} of performance response variable R_{ij} with desirable value is formulated, the evaluations of total probability P_t of candidate and the ranking of the multi-objective optimization can be performed according to the common procedure of the probability-based multiple objective optimization [6].

5.3 Applications

(1) Maximizing Yield with Constraints of Viscosity and Molecular Weight

A maximizing yield optimization with constraints of viscosity and molecular weight problem was proposed by Montgomery et al. [7], in which two input variables are involved, i.e., reaction time x_1 and temperature x_2; it contained three responses variables, i.e., the yield y_1 (%), the viscosity y_2 (cSt), and the molecular weight y_3 (Mr.) of the product. The designed experiment together with the test data is cited and displayed in Table 5.1.

In this problem, the optimization is to achieve maximum yield y_1 with the constraints of viscosity y_2 and molecular weight y_3 being $62 \le y_2 \le 68$ cSt and $y_3 \le 3400$ Mr.

The utility can be obtained according to above constraint conditions, and the results are given in Table 5.2.

This optimal problem involves complex optimization for yield y_1 as a beneficial performance response index while viscosity y_2 and molecular weight y_3 as desirable performance response indexes comprehensively. Therefore, the assessment for the partial preferable probability of yield y_1 could be treated as beneficial type variable with the corresponding procedure [1], while the assessments for partial preferable probabilities of the viscosity y_2 and molecular weight y_3 should be conducted as the procedure which was developed in the last section for both one side desirable and one range desirable performance response problems, individually. The evaluated results

Table 5.1 Designed experiments and results for optimization of maximizing yield with constraints of viscosity and molecular weight

No	Reaction time, x_1/min	Temperature, x_2/°C	Yield, y_1/%	Viscosity, y_2/cSt	Molecular weight, y_3/Mr.
1	80	76.67	76.5	62	2940
2	80	82.22	77	60	3470
3	90	76.67	78	66	3680
4	90	82.22	79.5	59	3890
5	85	79.44	79.9	72	3480
6	85	79.44	80.3	69	3200
7	85	79.44	80	68	3410
8	85	79.44	79.7	70	3290
9	85	79.44	79.8	71	3500
10	92.07	79.44	78.4	68	3360
11	77.93	79.44	75.6	71	3020
12	85	83.37	78.5	58	3630
13	85	75.52	77	57	3150

Table 5.2 Utility of the designed experiments under constraints of viscosity and molecular weight with optimization of maximizing yield

No	Yield, y_1/%	Viscosity, y_2/cSt	Molecular weight, y_3/Mr.	Utility for yield, y_1/%	Utility for viscosity	Utility for molecular weight
1	76.5	62	2940	76.5	1	1
2	77	60	3470	77	0	0
3	78	66	3680	78	1	0
4	79.5	59	3890	79.5	0	0
5	79.9	72	3480	79.9	0	0
6	80.3	69	3200	80.3	0	1
7	80	68	3410	80	1	0
8	79.7	70	3290	79.7	0	1
9	79.8	71	3500	79.8	0	0
10	78.4	68	3360	78.4	1	1
11	75.6	71	3020	75.6	0	1
12	78.5	58	3630	78.5	0	0
13	77	57	3150	77	0	1

of partial preferable probabilities P_{y1}, P_{y2}, P_{y3}, and total preferable probabilities P_t of this problem are given in Table 5.3.

It can be seen from Table 5.3 that the test scheme *No.* 10 exhibits the maximum total preferable probability at first glance, so the optimal status might be around test scheme *No.* 10.

Table 5.3 Assessed results of partial and total preferable probabilities for the chemical experiment

No	Response variables			Preferable probability			
	y_1/%	y_2/cSt	y_3/Mr.	P_{y1}	P_{y2}	P_{y3}	$P_t \times 10^3$
1	76.5	62	2940	0.0750	0.25	1/6	3.125
2	77	60	3470	0.0755	0	0	0
3	78	66	3680	0.0765	0.25	0	0
4	79.5	59	3890	0.0779	0	0	0
5	79.9	72	3480	0.0783	0	0	0
6	80.3	69	3200	0.0787	0	1/6	0
7	80	68	3410	0.0784	0.25	0	0
8	79.7	70	3290	0.0781	0	1/6	0
9	79.8	71	3500	0.0782	0	0	0
10	78.4	68	3360	0.0768	0.25	1/6	3.2
11	75.6	71	3020	0.0741	0	1/6	0
12	78.5	58	3630	0.0769	0	0	0
13	77	57	3150	0.0755	0	1/6	0

The yield of No. 10 is $y_1 = 78.4\%$, the viscosity variable is $y_2 = 68$ cSt, and the molecular weight is $y_3 = 3360$ Mr. at $x_1 = 92.07$ min and $x_2 = 79.44$ °C. All constraint conditions are met in the viewpoint of probability theory.

(2) Optimization of Maximizing Conversion Rate with Constraints of Desirable Thermal Activity

A problem of optimization of maximizing conversion rate with constraints of desirable thermal activity was studied by Myers [8]. Three input variables are considered in the experiment, which includes the reaction time x_1, temperature x_2, and percentage of catalyst x_3; two desirable performance response variables are involved, i.e., conversion rate Y_1 (%) and thermal activity Y_2 (W $s^{0.5}$/(m^2 K)). The designed experiment together with the test data is cited and displayed in Table 5.4.

As to this problem, the optimization also involves complex responses of conversion rate Y_1 (%) as a beneficial type of performance index and the thermal activity y_2 as desirable type of performance index by $50 \le Y_2 \le 65$ W $s^{0.5}$/(m^2 K) and as

Table 5.4 Designed experiments and results for optimization of maximizing conversion rate with constraints of desirable thermal activity

No	Input variables		Response variables		
	Reaction time x_1/min	Temperature x_2/°C	Catalyst x_3/%	Conversion rate Y_1/%	Thermal activity Y_2/ W $s^{0.5}$/(m^2 K)
1	45	48	0.682	74	53.2
2	55	48	0.682	51	62.9
3	45	58	0.682	88	53.4
4	55	58	0.682	70	62.6
5	45	48	2.682	71	57.3
6	55	48	2.682	90	67.9
7	45	58	2.682	66	59.8
8	55	58	2.682	97	67.8
9	41.59	53	1.682	76	59.1
10	58.41	53	1.682	79	65.9
11	50	44.59	1.682	85	60
12	50	61.41	1.682	97	60.7
13	50	53	0	55	57.4
14	50	53	3.364	81	63.2
15	50	53	1.682	81	59.2
16	50	53	1.682	75	60.4
17	50	53	1.682	76	59.1
18	50	53	1.682	83	60.6
19	50	53	1.682	80	60.8
20	50	53	1.682	91	58.9

Table 5.5 Assessed results of partial and total preferable probabilities for the maximizing conversion rate with constraints of desirable thermal activity

No	Response variables		Preferable probability		
	Y_1	Y_2	P_{Y1}	P_{Y2}	$P_t \times 10^3$
1	74	53.2	0.0473	0.0585	2.7648
2	51	62.9	0.0326	0.0584	1.9028
3	88	53.4	0.0562	0.0585	3.2879
4	70	62.6	0.0447	0.0585	2.6142
5	71	57.3	0.0453	0.0585	2.6527
6	90	67.9	0.0575	8.4559×10^{-5}	0.0049
7	66	59.8	0.0421	0.0585	2.4659
8	97	67.8	0.0619	0.0001	0.0064
9	76	59.1	0.0485	0.0585	2.8396
10	79	65.9	0.0504	0.0055	0.2772
11	85	60	0.0543	0.0585	3.1758
12	97	60.7	0.0619	0.0585	3.6242
13	55	57.4	0.0351	0.0585	2.0549
14	81	63.2	0.0517	0.0583	3.0139
15	81	59.2	0.0517	0.0585	3.0264
16	75	60.4	0.0479	0.0585	2.8022
17	76	59.1	0.0486	0.0585	2.8396
18	83	60.6	0.0531	0.0585	3.1011
19	80	60.8	0.0511	0.0585	2.9890
20	91	58.9	0.0581	0.0585	3.4000

close to 57.5 W $s^{0.5}/(m^2$ K) as possible. Therefore, the partial preferable probability for conversion rate y_1 is assessed by the common procedure with the corresponding procedure [1], and partial preferable probability of the thermal activity y_2 should be conducted by the procedure developed in the last section for one range desirability problem. The assessed results of partial and total preferable probabilities P_{y1}, P_{y2}, and P_t of this problem are shown in Table 5.5.

It can be seen from Table 5.5 that the test scheme *No.* 12 possesses the maximum total preferable probability at first glance, so the optimal status of this problem might be around test scheme *No.* 12.

Analogically, the data in Table 5.5 is regressed to get corresponding functions. The regressed function for the total preferable probability is

$$P_t \times 10^3 = -55.9337 + 2.3901x_1 - 0.1179x_2 + 7.0079x_3$$
$$-0.0250x_1^2 + 0.0010x_2^2 - 0.2800x_3^2 + 0.0019x_1x_2$$

$$
\begin{aligned}
&-0.0355x_2x_3-0.0893x_3x_1 \\
R^2 &= 0.8166.
\end{aligned}
\tag{5.7}
$$

P_t obtains its maximum value $P_{tmax} \times 10^3 = 3.657$ at $x_1 = 48.525$ min., $x_2 = 61.41$ °C, and $x_3 = 2.473\%$.

While, the regressed function for the conversion rate Y_1 is

$$
\begin{aligned}
Y_1 &= 497.1314 - 0.7916x_1 - 14.5965x_2 - 49.0071x_3 - 0.0733x_1^2 \\
&\quad + 0.1175x_2^2 - 5.1915x_3^2 + 0.0850x_1x_2 - 0.7750x_2x_3 + 2.2750x_3x_1 \\
R^2 &= 0.9199.
\end{aligned}
\tag{5.8}
$$

The conversion rate Y_1 gets its optimal value $Y_{1\mathrm{Opt.}} = 94.337\%$ at $x_1 = 48.525$ min., $x_2 = 61.41$ °C, and $x_3 = 2.473\%$.

The regressed function for the desirable thermal activity Y_2 is

$$
\begin{aligned}
Y_2 &= 73.4856 - 1.7884x_1 + 0.4035x_2 - 0.9000x_3 + 0.0334x_1^2 \\
&\quad + 0.0030x_2^2 + 0.0572x_3^2 - 0.0155x_1x_2 + 0.0625x_2x_3 - 0.0075x_3x_1 \\
R^2 &= 0.8918.
\end{aligned}
\tag{5.9}
$$

The desirable thermal activity Y_2 gets it optimal value $Y_{2\mathrm{Opt.}} = 57.545$ W $\mathrm{s}^{0.5}$/(m^2 K) at $x_1 = 48.525$ min., $x_2 = 61.41$ °C, and $x_3 = 2.473\%$.

These optimized consequences indicate that all the optimized responses are better than those of test scheme *No.* 12 of Table 5.5. Above optimal results meet the demands of the essence of the problem in the viewpoint of probability theory.

5.4 Concluding Remarks

The performance response variable with desirable value can be divided into types of "one side desirable value problem" and "one range desirable value problem"; the assessment of partial preferable probability of the corresponding type can be formulated in viewpoint of probability theory quantitatively; thereafter, the assessments for total preferable probability and ranking can be conducted according to the proposed procedure. The regression analysis for the total preferable probability and performance response variables could supply the optimal status of the multi-objective optimization problem with desirable response variable by letting the maximum of the total preferable probability properly.

References

1. M. Zheng, Y. Wang, H. Teng, A new "intersection" method for multi-objective optimization in material selection. Tehnički Glas. **15**(4), 562–568 (2021). https://doi.org/10.31803/tg-20210901142449
2. V.S. Galgali, M. Ramachandran, G.A. Vaidya, Multi objective optimal sizing of distributed generation by application of Taguchi desirability function analysis. SN Appl. Sci. **1**(742), 1–14 (2019). https://doi.org/10.1007/s42452-019-0738-3
3. G. Derringer, R. Suich, Simultaneous optimization of several response variables. J. Qual. Technol. **12**(4), 214–219 (1980). https://doi.org/10.1080/00224065.1980.11980968
4. T. Goyal, R.S. Walia, T.S. Sidhu, Taguchi and utility based concept for determining optimal process parameters of cold sprayed coatings for multiple responses. Int. J. Interact. Des. Manuf. **11**, 761–769 (2017). https://doi.org/10.1007/s12008-016-0359-7
5. P. Brémaud, *Probability Theory and Stochastic Processes, Universitext Series* (Springer, Cham, 2020), pp. 7–11. https://doi.org/10.1007/978-3-030-40183-2
6. M. Zheng, H. Teng, Y. Wang, Application of intersection method for multi-objective optimization in optimal test with desirable response variable. Tehnički Glas. **16**(2), 178–181(2022). https://doi.org/10.31803/tg-20211012135212
7. D.C. Montgomery, *Design and Analysis of Experiments*, 9th edn. (John Wiley & Sons, New Jersey, 2017), pp.500–511
8. R.H. Myers, D.C. Montgomery, C.M. Anderson-Cook, *Response Surface Methodology Process and Product Optimization Using Designed Experiments*, 3rd edn. (John Wiley & Sons, New Jersey, 2009), pp. 276–277

Chapter 6
Hybrids of Probability-Based Multi-objective Optimization with Experimental Design Methodologies

Abstract It describes the hybrids of probability-based multi-objective optimization with experiment design methodologies, including orthogonal experimental design, response surface design, and uniform experimental design. The total preferable probability of a candidate alternative is the unique decisive index for the alternative selection or optimization quantitatively. The optimization of multi-objective orthogonal experiment design is conducted by using range analysis to the total preferable probabilities comprehensively; the optimizations of multi-objective of response surface design and uniform experiment design are to get maximum of the total preferable probability integrally. Some application examples in materials selection are given.

6.1 Introduction

Usually in many industrial processes and experiments, quality improvement or optimization is conducted by using experimental design, such as orthogonal experimental design, response surface design, and uniform experimental design. Optimization for one individual objective separately could not give the appropriate consequence of the optimization for several objectives simultaneously in general, and the simultaneous optimization of the multi-objectives does not equal to any form of "superposition" of individual objective optimization actually.

Up to now, though several multi-objective optimization approaches have been proposed [1–5], the general mathematical treatment in these approaches is "additive" algorithm for the normalized evaluation indexes, and some methods even include personal factors. In the respect of probability theory, "additive" algorithm is not consistent with the essence of "simultaneous optimization of multiple indexes" [6]. In fact, if different normalization algorithms are applied, considerable differences in the results of these methods could be produced [7]. So, above discussion indicates

The original version of this chapter was revised: Equations 6.6, 6.10, 6.11, 6.12 and deleting the word "by" in the last line before Table 6.5 has been corrected. The correction to this chapter is available at https://doi.org/10.1007/978-981-99-3939-8_15

M. Zheng et al., *Probability-Based Multi-objective Optimization for Material Selection*, https://doi.org/10.1007/978-981-99-3939-8_6

that approach relying on any type of "additive" algorithm is at most semiquantitative methods in some sense.

As to optimization of multi-objective orthogonal experimental test design, Taguchi developed analysis methods, both "analysis of signal to noise ratio (SNR)" and "gray relational analysis (GRA)" are combined to solve the optimal problem [8]. The scaling factors, the non-equivalence of SNR for assessing beneficial and unbeneficial types of indicators, and target best type indicators as well, the "additive" algorithm and the personal factor in gray relational coefficient, etc., are all included in the treatment, which leads to the inevitable inherent shortcomings of this approach. Besides, the "comprehensive balance method" and "comprehensive scoring method" are also used to conduct the assessment of optimization of multi-objective orthogonal experimental design [8–10], which are not fully quantitative, but empirical ones instead.

Response surface methodology (RSM) is an integration of both statistical and mathematical technique, which is a useful optimal method. It is widely used in both formulation of new products and design improvement of existing product [11]. Pareto algorithm is usually used in response surface design for the optimization of multi-objective problem, but the inherent feature of uncertainty or "additive" algorithm remains [11].

Derringer et al. and Jorge et al. once proposed desirability function to transfer each response variable into a desirability value [1, 2], but this kind of approach is not coincident with the original idea of simultaneous optimization of multi-objective at all in the spirit of probability theory and away from the viewpoint of system theory.

Uniform experimental design methodology (UEDM) was developed by Fang and Wang, which is a novel experimental design method to meet the demand of very few amount of experiment number for valuable experiment, such as in missile design [12]. It has now been utilized in many fields with fruitful consequences and huge benefits. Similar to the optimization of multi-objective orthogonal test design, some treatments with "additive" algorithm and the personal factors are used to deal with its optimization of multiple objectives [13].

In the spirit of probability theory, "simultaneous optimization of multiple indexes" should adopt the form of "multiplication" algorithm for the partial probability of each independent event to get the joint probability of the "overall (integrated) event" appropriately [6]. Thus, one has to obtain the partial probability for each objective as an independent event in the multi-objective optimization process preliminarily, then the operation of probability method can be conducted accordingly.

In this chapter, the hybrids of probability-based multi-objective optimization (PMOO) with experimental design methodologies are conducted, including orthogonal experimental design, response surface design, and uniform experimental design.

6.2 Hybrid of Probability-Based Multi-objective Optimization with Orthogonal Experimental Design

6.2.1 Algorithm of the Hybrid for PMOO with Orthogonal Experimental Design

As stated in Chap. 3 that the simultaneous optimization of multi-objective is to transfer it into a single-objective optimization one by using the total preferable probability of a candidate alternative, and the total preferable probability is the unique and overall decisive index for the simultaneous optimization of multi-object orthogonal experimental design in respect of probability theory, therefore, range analysis algorithm in the general orthogonal experimental design for single objective can be performed for the total preferable probability unaffectedly [13].

The appropriate configuration of the experimental variables is thus corresponding to the optimized total preferable probability.

Till now, the combination of the probability-based method for simultaneous optimization of multi-objective with orthogonal experimental design is regulated.

6.2.2 Application of the Hybrid of PMOO with Orthogonal Experimental Design in Material Selection

(1) **Multi-objective optimization of molding plastics process for storage box with orthogonal experimental design**

The multi-objective optimization of molding plastics process for storage box with orthogonal experimental design and CAE software was conducted by Zhu et al. [10], which involves five independent factors, i.e., mold temperature A, melt temperature B, pressurizing time C, packing pressure D, and injection time E. Four levels are used in the orthogonal experimental design with moldflow. The buckling deformation (W) and the volume shrink mark index (S) are taken as the indexes of the multiple targets in this optimization problem [10].

Both the buckling deformation and the volume shrink mark indexes are all unbeneficial type of performance ones in this optimal problem.

The orthogonal experimental design for the molding plastics process of the storage box is given in Table 6.1 [10]. Shown in Table 6.2 are the experiment results and the evaluated preferable probabilities of this orthogonal experiment design for the molding plastics process.

Table 6.2 shows the maximum of total preferable probability P_i being test *No. 4*, which can be chosen as the proper configuration from multi-objective orthogonal design in the first glance directly.

Table 6.1 Independent variables and levels of the orthogonal experiment design for molding plastics [10]

Level	Independent factors				
	Mold temperature A (°C)	Melt temperature B (°C)	Pressurizing time C (s)	Packing pressure D (MPa)	Injection time E (s)
1	30	230	8	50	3.5
2	40	240	10	60	4.0
3	50	250	12	70	4.5
4	60	260	14	80	5.0

Table 6.2 Experiment results of the orthogonal design of molding plastics process

Test No.	Independent factors					Objective performance index		Partial preferable probability		Total preferable probability	Rank
	A	B	C	D	E	W (mm)	S (%)	P_{ij} for W	P_{ij} for S	$P_i \times 10^3$	
1	1	1	1	1	1	4.177	2.009	0.0552	0.0498	2.7469	12
2	1	2	2	2	2	3.701	1.732	0.0606	0.0587	3.5541	8
3	1	3	3	3	3	1.560	0.765	0.0847	0.0898	7.6099	2
4	1	4	4	4	4	0.807	0.637	0.0932	0.0939	8.7566	1
5	2	1	2	3	4	3.432	1.348	0.0636	0.0710	4.5181	7
6	2	2	1	4	3	4.449	1.590	0.0521	0.0633	3.2966	10
7	2	3	4	1	2	1.857	1.200	0.0814	0.0758	6.1689	3
8	2	4	3	2	1	3.639	2.178	0.0613	0.0443	2.7161	13
9	3	1	3	4	2	2.882	1.042	0.0698	0.0809	5.6466	4
10	3	2	4	3	1	2.546	1.225	0.0736	0.0750	5.5201	6
11	3	3	1	2	4	4.468	2.193	0.0519	0.0439	2.2761	14
12	3	4	2	1	3	3.864	2.919	0.0587	0.0205	1.2035	15
13	4	1	4	2	3	2.506	1.170	0.0741	0.0768	5.6850	5
14	4	2	3	1	4	3.475	1.930	0.0631	0.0523	3.3018	9
15	4	3	2	4	1	4.850	2.065	0.0476	0.0480	2.2829	11
16	4	4	1	3	2	8.258	1.812	0.0091	0.0561	0.5112	16

Furthermore, Table 6.3 displays the assessed results of range analysis for the total preferable probabilities of the orthogonal experiment design for the molding plastics process.

The result of range analysis in Table 6.3 shows that the order of the independent factors decreases in impact from C, A, D, E to B. The optimal configuration is $C_4A_1D_4E_4B_1$, while the CAE modeling experiment indicates that the corresponding buckling deformation and volume shrink mark index are 0.7323 mm and 0.4241%

Table 6.3 Assessed results of range analysis for the total preferable probabilities of the orthogonal experiment design for the molding plastics process

Factor	*A*	*B*	*C*	*D*	*E*
Level 1	5.6669	4.6492	2.2077	3.3553	3.3165
Level 2	4.1749	3.9182	2.8897	3.5578	3.9702
Level 3	3.6616	4.5845	4.8186	4.5398	4.4488
Level 4	2.9452	3.2969	6.5327	4.9957	4.7132
Range	2.7217	1.3523	4.3250	1.6404	1.3967
Order	2	5	1	3	4

[10], individually, which are much smaller than the minimum values of 0.8069 mm and 0.6370% of the results of test *No. 4* in the orthogonal experiment design in Table 6.2.

(2) **Multi-objective optimization of strengthening plate for automobile body with orthogonal experimental design in drawing process**

The problems of crack and wrinkle of strengthening steel B280VK plate with the thickness of 1.2 mm for automobile body in drawing process were studied by Gou et al. with orthogonal experiment design [14]. The evaluated objectives include wrinkle evaluation function Φ_1 and crack evaluation function Φ_2, and the blank holding force F (A), friction coefficient μ (B), and resistance coefficients C and D for draw beads loads P_1 and P_2 were taken as independent input variables. Orthogonal experimental design was used to conduct the optimal design [14]. The results of the strengthening plate for automobile body in drawing process are cited in Table 6.4.

The wrinkle evaluation function Φ_1 and crack evaluation function Φ_2 are unbeneficial type of performance indexes to the technique optimization.

Table 6.4 Results of the strengthening plate for automobile body in drawing process

No.	Independent input variable				Objective	
	A	*B*	*C*	*D*	Φ_1	Φ_2
1	160	0.15	0.05	0.40	0.132	0.943
2	150	0.15	0.15	0.30	0.120	0.898
3	140	0.18	0.15	0.40	0.138	1.103
4	160	0.12	0.15	0.35	0.129	0.824
5	140	0.12	0.05	0.30	0.114	0.833
6	160	0.18	0.10	0.30	0.131	3.420
7	140	0.15	0.10	0.35	0.134	0.887
8	150	0.12	0.10	0.40	0.142	0.794
9	150	0.18	0.05	0.35	0.122	1.202

The assessments of the partial and total preferable probabilities for the wrinkle evaluation function Φ_1 and crack evaluation function Φ_2 of this orthogonal experiment design are given in Table 6.5.

Table 6.5 indicates that test *No. 5* exhibits the maximum of the total preferable probability P_i, which can be chosen as one of the proper configuration of the multi-objective orthogonal design directly.

Table 6.6 presents the assessments of range analysis of the data of total preferable probabilities of Table 6.5.

The result of range analysis in Table 6.6 shows that the order of the independent input variables for impact decreases from *B*, *C*, *A* to *D*. The optimal configuration is $B_1C_1A_1D_2$, which is identical with the result of the complex comprehensive balance method accidentally [14].

Table 6.5 Assessments of the partial and total preferable probabilities for the wrinkle evaluation function Φ_1 and crack evaluation function Φ_2

No.	Partial preferable probability		Total	
	Φ_1	Φ_2	$P_i \times 10^2$	Rank
1	0.1086	0.1210	1.3144	6
2	0.1191	0.1227	1.4614	2
3	0.1033	0.1151	1.1896	9
4	0.1112	0.1255	1.3952	3
5	0.1243	0.1251	1.5558	1
6	0.1095	0.0294	0.3217	4
7	0.1068	0.1231	1.3154	5
8	0.0998	0.1266	1.2635	8
9	0.1173	0.1115	1.3080	7

Table 6.6 Assessments of range analysis of the data of total preferable probabilities

Level	Independent input variable			
	A	*B*	*C*	*D*
Level 1	1.3536	1.4048	1.3927	1.1129
Level 2	1.3442	1.3637	0.9668	1.3395
Level 3	1.0104	0.9397	1.3487	1.2559
Range	0.3432	0.4651	0.4259	0.2265
Order	3	1	2	4

6.3 Hybrid of Probability-Based Multi-objective Optimization with Response Surface Methodology Design

6.3.1 *Algorithm of the Hybrid for PMOO with Response Surface Methodology (RSM)*

In the simultaneous optimization of multiple responses, several responses are involved [15]. Let us take each response of the alternative as one objective of PMOO. Then some utilities of response might belong to the beneficial type, but other utilities of response might be attributed to unbeneficial type. Therefore, each utility of response makes its contribution to one partial preferable probability in linear manner according to its actual type, respectively.

Besides, as the alternative is an integral body of both beneficial and unbeneficial types of performance utility indexes, the overall/total preferable probability of an alternative can be gained by the product of all partial preferable probabilities for the simultaneous optimization of multiple responses in the spirit of probability theory. Through this procedure, the total/overall preferable probability thus transfers the multi-response optimization problem into a single response one, which is the uniquely decisive index of the alternative in the optimization.

Moreover, regression analysis is employed for the overall/total preferable probabilities of all alternatives of the designed experiment to gain a regressed function of the overall/total preferable probability. Thereafter, the maximum value of the overall/ total preferable probability and values of corresponding specific independent input variables are obtained by the usual algorithm of mathematics. The subsequent step is to regress each response to obtain its regressed function and then substitute the values of corresponding specific independent input variables into each regressed function of the response to gain its compromised result.

Till now, the procedure of hybrid of PMOO with response surface methodology design is well developed.

6.3.2 *Application of the Hybrid of PMOO with Response Surface Methodology Design in Material Selection*

(1) **Optimal design of PP/EPDM/GnPs/GF composites**

RSM was employed by Niyaraki, et al. to optimize the mechanical properties of impact strength and elastic modulus of polypropylene (PP)/ethylene propylene dine monomer (EPDM)/grapheme nanosheets (GnPs)/glass fiber (GF) hybrid nanocomposites with Box–Behnken method [16]. There are two responses of the optimization of the nanocomposite, i.e., impact strength and elastic modulus.

Three levels are applied for the three independent input variables, i.e., EPDM (5%, 10%, and 15 wt.%), GnPs (0, 1 and 2 wt.%), and glass fiber (10, 20 and 30 wt.%) [16].

Here the hybrid of PMOO with response surface methodology is utilized to reanalyze this problem.

Table 6.7 displays the experimental results of impact strength and elastic modulus, together with their partial preferable probabilities and total probabilities; both impact strength and elastic modulus are beneficial type of performance utility indexes.

Table 6.7 indicates that the alternative *No. 12* exhibits the maximum of the overall/total preferable probability P_i, which can be the proper chosen primarily at the first glance directly.

Furthermore, regression analysis is performed to the total preferable probability. The regressed function for the total preferable probability is

$$\begin{aligned} P_i \times 10^3 &= 0.5751 + 259.6483X_1 + 13.0371X_3 - 5592.7607X_1^2 \\ &+ 46.5832X_3^2 - 476.762X_1X_2 + 7.0495X_2X_3 \\ &- 167.1209X_1X_3 \end{aligned}$$

$$R^2 = 0.9931 \tag{6.1}$$

The function $P_i \times 10^3$ gets to its maximum value $P_{i\,\max} \times 10^3 = 6.2262$ at specific values of the independent input variables $X_1 = 0.0082$ wt.%, $X_2 = 0.3$ wt.%, and $X_3 = 0.15$ wt.%.

Simultaneously, the predicted values for impact strength and elastic modulus can be gained by substituting the above specific values of independent input variables X_1, X_2, and X_3 into the regression functions of impact strength and elastic modulus, individually.

The regressed function of impact strength is

$$\begin{aligned} f_{Im} &= 18.5833 + 6333.333X_1 + 601.6667X_2 + 178.3333X_3 \\ &- 154167X_1^2 - 691.667X_2^2 + 1233.333X_3^2 - 12500X_1X_2 \\ &- 100X_2X_3 - 2000X_1X_3, \end{aligned}$$

$$R^2 = 0.9958 \tag{6.2}$$

The predicted optimal value of impact strength is 195.19 J/m at $X_1 = 0.0082$ wt.%, $X_2 = 0.3$ wt.% and $X_3 = 0.15$ wt.% from Eq. (6.2).

The regression function of elastic modulus is

$$\begin{aligned} f_{El} &= 579.875 + 14025X_1 - 155X_2 - 710X_3 - 171250X_1^2 \\ &+ 2087.5X_2^2 + 2350X_3^2 - 23750X_1X_2 + 650X_2X_3 \\ &- 23750X_1X_3, \end{aligned}$$

Table 6.7 Test results of impact strength and elastic modulus, and their partial preferable probabilities and total preferable probabilities

Alternative	Independent input variable			Response	
	X_1 GnPs (wt.%)	X_2 glass fiber (wt.%)	X_3 EPDM (wt.%)	Elastic strength modulus values (MPa)	Impact values (J/m)
1	0	0.1	0.1	540	103
2	0.02	0.1	0.1	660	138
3	0	0.3	0.1	695	163
4	0.02	0.3	0.1	720	148
5	0	0.2	0.05	612	122
6	0.02	0.2	0.05	703	136
7	0	0.2	0.15	598	162
8	0.02	0.2	0.15	642	172
9	0.01	0.1	0.05	648	116
10	0.01	0.3	0.05	737	158
11	0.01	0.1	0.15	610	156
12	0.01	0.3	0.15	712	196
13	0.01	0.2	0.1	650	158
14	0.01	0.2	0.1	645	160
15	0.01	0.2	0.1	655	163

Alternative	Preferable probability		
	P_{ij} for elastic modulus	P_{ij} for impact strength	$P_i \times 10^3$
1	0.0550	0.0458	2.5144
2	0.0672	0.0613	4.1174
3	0.0707	0.0724	5.1212
4	0.0733	0.0657	4.8172
5	0.0623	0.0542	3.3753
6	0.0715	0.0604	4.3221
7	0.0609	0.0720	4.3795
8	0.0653	0.0764	4.9919
9	0.0659	0.0515	3.3981
10	0.0750	0.0702	5.2641
11	0.0621	0.0693	4.3019
12	0.0725	0.0871	6.3087
13	0.0661	0.0702	4.6427
14	0.0656	0.0711	4.6653
15	0.0667	0.0724	4.8265

$$R^2 = 0.9961. \tag{6.3}$$

The predicted optimal value for elastic modulus is 712.73 MPa at $X_1 = 0.0082$ wt.%, $X_2 = 0.3$ wt.% and $X_3 = 0.15$ wt.% from Eq. (6.3).

The tested result is 195.17 J/m for impact strength and 713.08 MPa for elastic modulus [16], which agrees with the predicted optimal data very well and also not far from the experimental results of the test *No. 12* of Table 6.7.

(2) **Maximizing yield and minimizing molecular weight with desired viscosity**

A simultaneous optimal problem of maximizing yield and minimizing molecular weight with desired viscosity was once studied by Myers et al., which involves two input variables reaction time x_1 and temperature x_2 [11]. There are three responses variables, i.e., the yield y_1 (%), the viscosity y_2 (cSt), and the molecular weight y_3 (Mr.) of the product. The relevant data are given in Table 6.8.

The viscosity y_2 (cSt) is with desired value of 65 cSt [11], so the utility U_{i2} of response y_{i2} of the actual experiment result could be indicated by its actual deviation from desired value of 65 cSt, i.e.,

$$U_{i2} = |y_{i2} - 65|. \tag{6.4}$$

In Eq. (6.4), i expresses the number of the experiment or alternative. In such case, the utility U_{i2} has the characteristics of "the lower the better", which belongs to

Table 6.8 Designed experiment and results of maximizing yield and minimizing molecular weight with desired viscosity

No	Independent input variable		Response			
	Reaction time, x_1/min	Temp., x_2/°C	Yield, y_1/%	Viscosity, y_2/ cSt	$U_{i2} = \| (y_{i2} - 65) \|$	Molecular weight, y_3/ Mr.
1	80	76.67	76.5	62	3	2940
2	80	82.22	77	60	5	3470
3	90	76.67	78	66	1	3680
4	90	82.22	79.5	59	6	3890
5	85	79.44	79.9	72	7	3480
6	85	79.44	80.3	69	4	3200
7	85	79.44	80	68	3	3410
8	85	79.44	79.7	70	5	3290
9	85	79.44	79.8	71	6	3500
10	92.07	79.44	78.4	68	3	3360
11	77.93	79.44	75.6	71	6	3020
12	85	83.37	78.5	58	7	3630
13	85	75.52	77	57	8	3150

unbeneficial type of performance index. Therefore, the assessment of partial preferable probability for desired yield y_{i2} is conducted by using its utility U_{i2} as an unbeneficial type of performance index as stated in Chap. 3.

The assessment of partial preferable probability for maximizing yield y_1 is conducted according to the usual procedures of the PMOO as beneficial type of performance index and minimizing molecular weight y_3 as unbeneficial type of performance index.

The assessments for partial and total preferable probabilities P_{y1}, P_{y2}, P_{y3}, and P_i of this product experiment are displayed in Table 6.9.

The data in Table 6.9 shows that the test *No. 3* is with the maximum total preferable probability, followed by *No. 1*, *No. 7*, and *No. 10.*

Furthermore, regression of the total preferable probability P_i can be done to gain more accurate optimization. The regressed result for the total preferable probability is

$$\begin{aligned} P_i \times 10^3 = & -203375.1310 - 2654.9450x_1 + 215.4526x_2 \\ & + 15.5210x_1^2 - 2.6907x_2^2 - 0.0038x_1x_2 \\ & + 75610.0916\ln(x_1) - 0.0403x_1^3 + 0.0112x_2^3 \end{aligned}$$

$$R^2 = 0.7675 \tag{6.5}$$

P_i gains its maximum value $P_{i\,max} \times 10^3 = 0.9334$ at $x_1 = 91.0622$ min and $x_2 = 77.6053$ °C.

Table 6.9 Assessments for partial and total preferable probabilities of the desired viscosity

No.	Response			Preferable probability				Rank
	y_1/%	y_2/cSt	y_3/Mr	P_{y1ij}	P_{y2ij}	P_{y3ij}	$P_i \times 10^3$	
1	76.5	62	2940	0.0750	0.1132	0.0869	0.7376	2
2	77	60	3470	0.0755	0.0755	0.0751	0.4275	7
3	78	66	3680	0.0765	0.1509	0.0704	0.8120	1
4	79.5	59	3890	0.0779	0.0566	0.0657	0.2897	10
5	79.9	72	3480	0.0783	0.0377	0.0748	0.2211	11
6	80.3	69	3200	0.0787	0.0943	0.0811	0.6021	5
7	80	68	3410	0.0784	0.1132	0.0764	0.6781	3
8	79.7	70	3290	0.0781	0.0755	0.0791	0.4662	6
9	79.8	71	3500	0.0782	0.0566	0.0744	0.3293	9
10	78.4	68	3360	0.0768	0.1132	0.0775	0.6743	4
11	75.6	71	3020	0.0741	0.0566	0.0851	0.3570	8
12	78.5	58	3630	0.0769	0.0377	0.0715	0.2075	12
13	77	57	3150	0.0755	0.0189	0.0822	0.1171	13

Meanwhile, the regressed result for the yield y_1 is

$$\begin{aligned} y_1 = & -326843.3710 - 4410.0970x_1 + 48.0746x_2 \\ & + 0.0180x_1x_2 + 25.9987x_1^2 - 0.4803x_2^2 \\ & + 124747.8522\ln(x_1) + 0.0682x_1^3 + 0.0014x_2^3, \end{aligned}$$

$$R^2 = 0.9926 \tag{6.6}$$

The yield y_1 gains its proper value of $y_{1\text{Opt.}} = 78.2722\%$ at $x_1 = 91.0622$ min and $x_2 = 77.6053$ °C.

Simultaneously, the regressed result for viscosity y_2 is

$$\begin{aligned} y_2 = & 1454310.2050 + 20242.9592x_1 + 2436.0490x_2 \\ & - 0.0900x_1x_2 + 117.4347x_1^2 - 29.7790x_2^2 \\ & - 580304\ln(x_1) + 0.3025x_1^3 + 0.1215x_2^3, \end{aligned}$$

$$R^2 = 0.9723 \tag{6.7}$$

The viscosity y_2 obtains its appropriate value $y_{2\text{Opt.}} = 68.8928$ cSt at $x_1 = 91.0622$ min and $x_2 = 77.6053$ °C.

Subsequently, the regressed result for molecular weight y_3 is

$$\begin{aligned} y_3 = & -176217474.000 - 2438309.5500x_1 - 13078.9964x_2 \\ & - 5.7600x_1x_2 + 14549.0282x_1^2 + 170.7936x_2^2 \\ & + 68063495.1500\ln(x_1) - 8.5337x_1^3 - 0.7128x_2^3 \end{aligned}$$

$$R^2 = 0.9238 \tag{6.8}$$

The optimal molecular weight y_3 gains it proper value $y_{3\text{Opt.}} = 3590.0681$ Mr. at $x_1 = 91.0622$ min and $x_2 = 77.6053$ °C.

Evidently, the optimal status of this problem is not far from test *No. 3* of Table 6.8.

6.4 Hybrid of Probability-Based Multi-objective Optimization with Uniform Experimental Design Methodology

6.4.1 Algorithm of the Hybrid for PMOO with Uniform Experimental Design Methodology (UED)

The optimizations of multi-objective of uniform experimental design is to get maximum of the total preferable probability integrally for the simultaneous optimization of multiple responses. Similar to the optimization problem of multiple responses, each response of the alternative is one objective of PMOO analogically, and each utility of response is divided into beneficial or unbeneficial type according to its function. Every utility contributes to one partial preferable probability in linear manner according to its actual type, respectively.

Thereafter, the overall/total preferable probability of an alternative can be gained by the product of all partial preferable probabilities for the simultaneous optimization of multi-response in the respect of probability theory. Moreover, regression analysis is employed for the overall/total preferable probabilities of all alternatives of the designed experiment to gain a regressed function of the overall/total preferable probability, and the maximum of overall/total preferable probability corresponds to the optimal status of the problem.

The following step is to regress each response to obtain its regressed function and then substitute the values of corresponding specific independent input variables into each regressed function of the response to gain its compromised result.

Till now, the procedure of hybrid of PMOO with uniform experimental design is well constructed.

6.4.2 Application of the Hybrid of PMOO with Uniform Experimental Design Methodology in Material Selection

(1) **Multi-objective optimization of injection molding process parameters with uniform experimental design**

A multi-objective optimization of injection molding process parameters by using TOPSIS and UED was conducted with moldflow simulation by Cheng et al. [17]. There are three objective indicators, i.e., volume shrink rate ϕ_{shr} (%), sink index I_{sink} (%), and buckling deformation W (mm); the input variables are melt temperature X_1, injection time X_2, and packing pressure X_3, while retaining injection pressure of 100 MPa, packing time of 3 s, and the cooling time of 5 s as invariables; the uniform experimental design was then performed.

The uniform experimental design table for this multi-objective optimization is $U_{10}(5^3)$ [17]. The levels of variables are cited in Table 6.10. The results of this multi-objective optimization and their injection molding process parameters are given in Table 6.11.

The volume shrink rate ϕ_{shr}, sink index I_{sink}, and buckling deformation W can be taken as their utilities directly, and they belong to unbeneficial type of performance indicators for this optimization [13].

The partial and total preferable probabilities for the volume shrink rate ϕ_{shr}, sink index I_{sink}, and buckling deformation W are given in Table 6.12.

From Table 6.12, it can be seen that the test *No. 2* exhibits the maximum of the total preferable probability P_i closely followed by test *No. 1*, so they could be chosen directly as one of the optimal configuration at the first glance in this multi-objective optimization with uniform experimental design.

The test result at pack pressure $P = 90$ MPa, melt temperature $\theta_{melt} = 275$ °C and injection time = 0.7 s, was shown by Cheng, et al., which is the compromised combination of the parameters of the tests *No. 1* and *No. 2* [17], and shows smaller volume shrink rate ϕ_{shr} of 6.027%, sink index I_{sink} of 3.123%, and buckling deformation W of 0.386 mm, indicating the validity of the new evaluations.

Furthermore, regression of the total preferable probability can be done to gain more accurate optimization. The regressed result for the total preferable probability

Table 6.10 Levels of variables for this multi-objective optimization

Factors	Levels				
Label	1	2	3	4	5
Melt temperature, X_1 (°C)	265	270	275	280	285
Injection time, X_2 (s)	0.50	0.55	0.60	0.65	0.70
Packing pressure, X_3 (MPa)	70	75	80	85	90

Table 6.11 Results of this multi-objective optimization and their injection molding process parameters

No.	X_1 (°C)	X_2 (s)	X_3 (MPa)	ϕ_{shr} (%)	I_{sink} (%)	W (mm)
1	270	0.7	90	6.037	3.230	0.4048
2	280	0.5	90	6.200	3.393	0.3735
3	285	0.55	75	6.895	4.057	0.4306
4	280	0.7	70	6.937	4.090	0.4695
5	275	0.6	80	6.587	3.744	0.4314
6	270	0.5	70	7.114	4.286	0.5103
7	285	0.65	85	6.299	3.471	0.3815
8	275	0.6	80	6.587	3.744	0.4314
9	265	0.55	85	6.463	3.614	0.4406
10	265	0.65	75	6.777	3.938	0.5054

Table 6.12 Partial and total preferable probabilities for the volume shrink rate ϕ_{shr}, sink index I_{sink}, and buckling deformation W

No.	P_{ij} for ϕ	P_{ij} for I	P_{ij} for W	$P_i \times 10^3$	Rank
1	0.1084	0.1140	0.1074	1.3279	2
2	0.1059	0.1097	0.1144	1.3297	1
3	0.0953	0.0920	0.1016	0.8917	6
4	0.0947	0.0911	0.0929	0.8020	7
5	0.1000	0.1003	0.1015	1.0182	5
6	0.0921	0.0860	0.0838	0.6622	9
7	0.1045	0.1076	0.1126	1.2658	3
8	0.1000	0.1003	0.1016	1.0184	5
9	0.1019	0.1038	0.0994	1.0516	4
10	0.0971	0.0952	0.0849	0.7846	8

is

$$\begin{aligned} P_i \times 10^3 = {} & -258.3560 + 3.0225x_1 - 0.9225x_3 - 0.0107x_1^2 \\ & + 23.3592x_2^2 + 0.0095x_3^2 + 1.2700 \times 10^{-5}x_1^3 \\ & - 25.4000x_2^3 - 3.0000 \times 10^{-5}x_3^3, \end{aligned}$$

$$R^2 = 1. \tag{6.9}$$

P_i gains its maximum value $P_{i\,max} \times 10^3 = 1.5959$ at $x_1 = 280.84$ °C, $x_2 = 0.61$ s, and $x_3 = 90.00$ MPa.

Meanwhile, the regressed result for the yield ϕ_{shr} is

$$\begin{aligned} \phi_{shr} = {} & 119.5035 - 1.5099x_1 + 1.0815x_3 + 0.0052x_1^2 \\ & - 68.5306x_2^2 - 0.0073x_3^2 - 6 \times 10^{-6}x_1^3 + 74.6667x_2^3 \\ & + 2.0000 \times 10^{-6}x_3^3, \end{aligned}$$

$$R^2 = 1 \tag{6.10}$$

The volume shrink rate ϕ_{shr} (%) obtains its appropriate $\phi_{shr\,Opt.} = 5.449\%$ at $x_1 = 280.84$ °C, $x_2 = 0.61$ s, and $x_3 = 90.00$ MPa.

The regressed result for volume sink index I_{sink} is

$$\begin{aligned} I_{sink} = {} & -299.3830 + 3.0302x_1 + 1.1567x_3 - 0.0113x_1^2 \\ & - 50.7063x_2^2 - 0.0102x_3^2 + 1.4000 \times 10^{-5}x_1^3 \\ & + 55.0000x_2^3 + 2.2300 \times 10^{-5}x_3^3 \end{aligned}$$

$$R^2 = 1 \tag{6.11}$$

The volume sink index I_{sink} (%) obtains its appropriate $I_{\text{sink Opt.}} = 2.812\%$ at $x_1 = 280.84$ °C, $x_2 = 0.61$ s, and $x_3 = 90.00$ MPa.

The regressed result for buckling deformation W is

$$W = -62.3534 + 0.7395x_1 - 0.0428x_3 - 0.0028x_1^2 + 12.3638x_2^2 - 0.0008x_3^2 + 3.4300 \times 10^{-6}x_1^3 - 13.6333x_2^3 + 8.7700 \times 10^{-6}x_3^3,$$

$$R^2 = 1 \tag{6.12}$$

The buckling deformation W obtains its appropriate $W_{\text{Opt.}} = 0.491$ mm at $x_1 = 280.84$ °C, $x_2 = 0.61$ s, and $x_3 = 90.00$ MPa.

Comprehensively, the optimal results by using regression approach are successful.

(2) **Multi-objective optimization design for composition of diamond abrasive tools by means of uniform experimental design**

Composition design for diamond abrasive tools was conducted by means of uniform experimental design by Liu et al. [18]. Three variables are included, i.e., the contents of copper powder X_1 (volume faction), chromium oxide X_2 (volume faction), and zinc oxide X_3 (volume faction) [18, 19]. While the objective indicators contain the grinding ratio Y_1 of the diamond sample (%), the removal rate Y_2 (mm/s) and the grinding efficiency Y_3 (g/s) of the sample [18, 19], which could be taken as the utility indexes directly in the alternatives selection, are given in Table 6.13. The utility indexes of grinding ratio Y_1 and grinding efficiency Y_3 are attributed to beneficial type of indexes, and the utility index of removal rate Y_2 is attributed to unbeneficial type of indicator in this optimization, respectively.

Table 6.14 shows the evaluation consequences of partial preferable probabilities and the overall preferable probabilities for each test alternative.

Table 6.13 Uniform experiment design $U_7(7^3)$ and their results for composition optimization of diamond abrasive tools

No.	Parameter and result of alternate					
	X_1	X_2	X_3	Y_1 (%)	Y_2 (mm/s)	Y_3 (g/s)
1	8	3	2	1.542	0.214	0.106
2	10	6	3	1.717	0.153	0.105
3	12	2	4	1.795	0.109	0.078
4	14	5	1.5	1.933	0.099	0.083
5	16	1	2.5	1.101	0.190	0.091
6	18	4	3.5	2.671	0.055	0.067
7	20	7	4.5	1.791	0.136	0.137

Table 6.14 Partial and overall favorable probabilities for the optimization of diamond abrasive tools

No.	Partial favorable probabilities			Overall	
	P_{Y1}	P_{Y2}	P_{Y3}	$P_i \times 10^3$	Rank
1	0.1229	0.0593	0.1589	1.1585	6
2	0.1368	0.1251	0.1574	2.6951	5
3	0.1430	0.1726	0.1169	2.8869	4
4	0.1540	0.1834	0.1244	3.5149	3
5	0.0877	0.0852	0.1364	1.0200	7
6	0.2128	0.2309	0.1004	4.9353	1
7	0.1427	0.1435	0.2054	4.2055	2

From Table 6.14, it can be seen that the test *No. 6* possesses the maximum of the overall preferable probability P_i in the ranking. Therefore, test *No. 6* can be chosen as one of the optimal alternatives directly at the first glance.

Furthermore, regression of the total preferable probability of this optimization can be done to obtain more accurate optimization. The regressed result for the total preferable probability P_i is

$$P_i \times 10^3 = -9.5710 + 0.8658x_1 + 2.3484x_2 + 0.1482x_3 - 0.0244x_1^2 - 0.2564x_2^2 + 0.0820x_3^2$$

$$R^2 = 1. \tag{6.13}$$

P_i obtains its maximum value $P_{i\,\max} \times 10^3 = 5.8275$ at $x_1 = 17.7663$, $x_2 = 4.5804$, and $x_3 = 4.5000$.

Meanwhile, the regressed result for Y_1 is

$$Y_1 = -2.2743 + 0.1356x_1 + 1.0791x_2 + 0.5143x_3 - 0.0029x_1^2 - 0.1287x_2^2 - 0.0554x_3^2,$$

$$R^2 = 1. \tag{6.14}$$

Y_1 obtains its proper value $Y_{1\,\text{opt.}} = 2.7199$ at $x_1 = 17.7663$, $x_2 = 4.5804$, and $x_3 = 4.5000$.

The regressed result for Y_2 is

$$Y_2 = 0.8136 - 0.0646x_1 - 0.0975x_2 - 0.0067x_3 + 0.0021x_1^2 + 0.0114x_2^2 - 0.0034x_3^2$$

$$R^2 = 1. \tag{6.15}$$

Y_2 obtains its proper value Y_2 opt. = 0.0196 at $x_1 = 17.7663$, $x_2 = 4.5804$, and $x_3 = 4.5000$.

The regressed result for Y_3 is

$$Y_3 = 0.3393 - 0.0238x_1 - 0.0312x_2 - 0.0286x_3 + 0.0008x_1^2 + 0.0044x_2^2 + 0.0043x_3^2$$

$$R^2 = 1. \tag{6.16}$$

Y_3 obtains its proper value Y_3 opt. = 0.0754 at $x_1 = 17.7663$, $x_2 = 4.5804$, and $x_3 = 4.5000$.

Obviously, the optimal results by using regression approach are wholly more superior to the test *No. 6*.

6.5 Conclusion

In above discussion, the probability theory-based multi-objective optimization is combined with orthogonal experimental design, response surface methodology, and uniform experimental design. Each utility index of material performance indicator contributes to a partial preferable probability in the assessment quantitatively; the total preferable probability of a candidate alternative is the product of all partial preferable probabilities, which thus naturally transfers the multi-objective problem into a single-objective one. Thereafter, as regard to hybrid of PMOO with orthogonal experiment design, the total preferable probabilities of all alternatives are employed to conduct range analysis and thus the multi-objective orthogonal test design comprehensively; with respect to hybrid of PMOO with response surface methodology and uniform experimental design, the total preferable probability of all alternatives is regressed to gain its maximum and the optimal status. The evident advantages and physical essence of the hybrids for multi-objective optimization of experiment designs are clear, which could supply a novel and simple way for multi-objective experimental designs.

References

1. G. Derringer, R. Suich, Simultaneous optimization of several response variables. J. Qual. Technol. **12**, 214–219 (1980). https://doi.org/10.1080/00224065.1980.11980968
2. L.R. Jorge, B.L. Yolanda, T. Diego, P.L. Mitzy, R.B. Ivan, Optimization of multiple response variables using the desirability function and a Bayesian predictive distribution. Res. Comput. Sci. **13**, 85–95 (2017)

3. S. Opricovic, G.H. Tzeng, Compromise solution by MCDM methods: a comparative analysis of VIKOR and TOPSIS. Eur. J. Oper. Res. **156**, 445–455 (2004). https://doi.org/10.1016/S0377-2217(03)00020-1
4. A. Shanian, O. Savadogo, TOPSIS multiple-criteria decision support analysis for material selection of metallic bipolar plates for polymer electrolyte fuel cell. J. Power Sources **159**, 1095–1104 (2006)
5. R. Kumar, J. Jagadish, A. Ray, Selection of material for optimal design using multi-criteria decision making. Procedia Mater. Sci. **6**, 590–596 (2014). https://doi.org/10.1016/j.mspro.2014.07.073
6. P. Brémaud, *Probability Theory and Stochastic Processes*, Universitext Series (Springer, Cham, 2020), pp. 7–11. https://doi.org/10.1007/978-3-030-40183-2
7. W. Yang, S. Chon, C. Choe, J. Yang, Materials selection method using TOPSIS with some popular normalization methods. Eng. Res. Express **3**, 015020 (2021). https://doi.org/10.1088/2631-8695/abd5a7
8. M. Teruo, *Taguchi Methods, Benefits, Impacts, Mathematics, Statistics, and Applications* (ASME Press, New York, 2011), pp.47–204
9. C. Obara, F.M. Mwema, J.N. Keraita et al., A multi-response optimization of the multi-directional forging process for aluminum 7075 alloy using grey-based Taguchi method. SN Appl. Sci. **3**(596), 1–20 (2021). https://doi.org/10.1007/s42452-021-04527-2
10. J. Zhu, W. Huang, Q. Zhang, et al., Multi-objective optimization of vehicle built-in storage box injection molding process parameters based on grey Taguchi method. Plastics Sci. Techno. **47**, 63–68 (2019). 005.3360(2019)04.0063-06
11. R.H. Myers, D.C. Montgomery, *Response Surface Methodology, Process and Product Optimization Using Designed Experiments*, 4th edn. (Wiley, New Jersey, 2016)
12. K.-T. Fang, M.-Q. Liu, H. Qin, Y.-D. Zhou, *Theory and Application of Uniform Experimental Designs* (Science Press & Springer, Beijing & Singapore, 2018). https://doi.org/10.1007/978-981-13-2041-5
13. M. Zheng, Y. Wang, H. Teng, A novel method based on probability theory for simultaneous optimization of multi-object orthogonal test design in material engineering. Kovove Mater. **60**(1), 45–53 (2022). https://doi.org/10.31577/km.2022.1.45
14. C. Gou, J. Dong, Multi-objective optimization on strengthening plate in automobile body during drawing process based on orthogonal test. Forg. Stamp. Technol. **43**, 41–44 (2018)
15. M. Zheng, Y. Wang, H. Teng, Hybrid of "Intersection" algorithm for multi-objective optimization with response surface methodology and its application. Technicki Glas. **16**(4), 454–457 (2022). https://doi.org/10.31803/tg-20210930051227
16. M.N. Niyaraki, F. A.Ghasemi, I. Ghasemi, S. Daneshpayeh, Predicting of impact strength and elastic modulus of polypropylene/EPDM/graphene/glass fiber nanocomposites by response surface methodology. Technicki Glas. **15**(2), 169–177 (2021). https://doi.org/10.31803/tg-20190204023624
17. J. Cheng, J. Tan, J. Yu, Multi-objective robust optimization of injection molding process parameters based on TOPSIS. J. Mech. Eng. **47**, 27–32 (2011). https://doi.org/10.3901/JME.2011.06.027
18. T. Liu, X. Ye, L. Huang, W. Zhou, Composition design for diamond abrasive tools by means of uniform experiment design, in *Proceedings of 2012 National Superhard Material Technology Development Forum*, Yichang, China, 24–25 Oct (2012), pp. 167–172
19. M. Zheng, Y. Wang, H. Teng, A novel approach based on probability theory for material selection. Materialwiss. Werkstofftechnik **53**(6), 666–674 (2022). https://doi.org/10.1002/mawe.202100226

Chapter 7
Discretization of Simplified Evaluation in Probability-Based Multi-objective Optimization by Means of GLP and Uniform Experimental Design

Abstract Discretization treatment of evaluation in probability-based multi-objective optimization by means of GLP and uniform experimental design is illuminated, which provides efficient simplification with low discrepancy. Some examples of application of multi-objective optimization with complicated integrals and estimation of extreme value are given, including evaluations of multi-objective optimization in structure and material designs for tower crane boom tie rods, optimal design for the composition of rubber, linear and nonlinear programming problems with domain in non-regular area, and multi-objective optimization of numerical control machining parameters for high efficiency and low carbon.

7.1 Introduction

Preliminarily, a proper approximation for estimating definite integral has continuously been a fantastic problem since the creation of integration, which is a useful algorithm for practical science and engineering, theoretical analysis, and information processing. In most practical conditions, the integrand is not always too simple to get its accurate solution or data, which induces the difficulty of evaluating it exactly, and thus, a proper approximation of a definite integral with certain precision is welcome for practical application. Therefore, it is valuable to find a proper approximation for a definite integral.

In the one variable (1 dimension, 1-D) case, there are many classical quadrature rules, such as the rectangle rule (midpoint rule), Simpson's rule, the trapezoidal rule, or the Gauss rule, which have the following basic form [1],

$$T_n(y) = \sum_{i=0}^{n} q_i y(x_i) \tag{7.1}$$

with the quadrature points $x_0, x_1, x_2, \ldots, x_i, \ldots, x_n$ being within range of [0, 1], and the weighting factors of $q_0, q_1, q_2, \ldots, q_i, \ldots, q_n$.

M. Zheng et al., *Probability-Based Multi-objective Optimization for Material Selection*, https://doi.org/10.1007/978-981-99-3939-8_7

As to the trapezoidal rule, $q_0 = q_n = 1/(2n)$, for other weights, $q_i = 1/n$ with $i = 1, 2, \ldots, n - 1$. If $f \in C^2([0; 1])$, the error of the trapezoidal rule is of the order $O(n^{-2})$.

Furthermore, in the case of s variables (s-D), it leads to the following expression

$$T_n(y) = \sum_{i=0}^{n} w_i y(x_i) \tag{7.2}$$

with the set of s-fold quadrature points $\{\boldsymbol{x}_0, \boldsymbol{x}_1, \boldsymbol{x}_2, \ldots, \boldsymbol{x}_i, \ldots, \boldsymbol{x}_n\}$ in the domain of $[0, 1]^s$. Thus, the total number of nodes is $N = (n + 1)^s$, which increases quickly with the dimension number s. While in terms of the practical number $N = n + 1$ of integration nodes, the error is of the order $O(N^{-2/s})$ [1].

As to higher dimensions, there exist some actual problems, such as error convergence. This phenomenon was called the "curse of dimensionality" in general [1].

Monte Carlo (MC) simulation was developed as an algorithm with stochastic sampling in mid-1940s. However, it needs a large number of random numbers (sampling points) in the simulation calculation, which is its inherent shortcoming besides the rather slow convergence speed [2, 3].

Early in 1959, Korobov proposed the idea of a point set with uniform distribution, thereafter Hua and Wang made the development of the good lattice point (GLP) method with low discrepancy in numerical integration in 1960s [4]. In the spirit of the GLP, the convergence speed of integration is much higher than the Monte Carlo method. In 1980s, Fang and Wang formulated a uniform experimental design method on the basis of the uniform sampling or "good lattice point" [2, 3]. In the uniform experimental design, the distribution of these sampling points is uniform in the space with well-deterministic positions instead of random spreads. Such kinds of algorithms are called the "quasi-Monte Carlo method" (QMC) thereafter [5–9].

However, the so-called curse of dimensionality problem perplexed the application of QMC method for many years till 1990s [5–9]. The dramatic change happened in 1990s when Paskov and Traub employed Halton sequences and Sobol sequences to account a ten-tranche Collateralized Mortgage Obligation (CMO) in supper space with very high dimensions even up to 360 dimensions, and they found that QMC methods exhibited very good convergence with respect to the convenient MC methods, and antithetic MC methods [5–9]. Afterward, a lot of similar phenomena appeared in various pricing problems by applying different types of low-discrepancy sequences [8, 11]. Papageorgiou et al. even reported that a simulation precision of order 10^{-2} was got with less than 500 points in 25-dimensional integral [11], and an empirical convergence rate approaching to n^{-1} was gained in their tests rather than the $n^{-1/2}$ of Monte Carlo.

The above results are actually counterintuitive, so it was difficult to comprehend the strange convergence speed of the point set with the low-discrepancy sequences superior to that of the random points. An idea of a so-called weighted discrepancy was developed by Sloan and Wozniakowski to explain this conundrum [7], while other concept of effective dimensions was proposed by Caflisch et al. to illuminate the miracle [10]. Anyhow, these consequences indicate the effectiveness of QMC methods though the reason behind is not clarified.

In fact in 1980s, Ripley once mentioned that the expectation of the mean square error of the sample could decrease with the spatial correlation of the samples in spatial sampling problem [12], and the expression for the sample mean $E(V_{\text{ran}})$ was further formulated by Dunn and Harrison as [13], $E(V_{\text{ran}}) = \frac{1-f}{n}\frac{N}{N-1}\left\{\sigma^2 - \overline{\text{cov}(Z_i, Z_j)^{ij}}\right\}$, in which N expresses the number observations in the population, n is the sample size, f indicates the finite population correction (n/N), σ^2 stands the population variance and $\overline{\text{cov}(Z_i, Z_j)^{ij}}$ presents the average covariance between all possible pairs of sampling points in the population. This expression reflects that the actual number of sampling required could decrease with the spatial correlation of the samples. This might be related to the above-mentioned counterintuitive phenomena by using QMC in high dimensions in the previous paragraph. The technique of spatial sampling with spatial correlation of the samples has been frequently employed in spatial statistics and geography with success these years [14].

Actually, the integrand of an integral is with its definite form and explicit physical meaning in general. So, the value of the integrand evolutes according to a definite rule with the point in space from one position to another, and therefore, it is more appropriate to perform the numerical integration according to a point set which is with a definite rule and has a regular distribution in the relevant space in principle.

Here in this chapter, we try to adopt a certain number of sampling points with regular distribution to conduct approximate evaluation for a definite integral first. It aims to give an efficient approach with certain precision for a definite integral. The feature analysis of a periodic function in its monoperiod is performed primarily. The consequences indicate that 11 sampling points in one period could provide an efficient approximation to its peak value with less error or deviation, which promotes the exploration on the use of 11 sampling points to carry out an efficient approach for the definite integral within its monotonic peak domain. Thereafter, a similar analysis for two and three variables issues is conducted as well. Moreover, typical examples of definite integral for some physical problems are supplied to display the reasonability of the approach. Furthermore, discretization of complicated integral in the probability-based multi-objective optimization is conducted, of which the distribution of sampling points follows the rules of GLP and uniform experimental design.

7.2 Fundamental Characteristics of Uniform Experimental Design

In nowadays, the application of uniform experimental design is increasing significantly, ranging from Chinese Missile Design and medicine design to Ford Motor Co. Ltd of USA for the automotive engine design and the standard practices for computer experiments to support the early stage of the production design [15].

7.2.1 *Main Features of Uniform Experimental Design*

Fang and Wang proposed the uniform experimental design (UED) in 1980s [2–4], the main features of UED include:

(A) **Uniformity**

The sampling points are uniformly distributed in the variable space; therefore, it gets other name "space filling design" sometimes in the literatures. UED arranges the experiment design by using a "uniform design table" without any randomness.

(B) **Overall Mean Model**

UED expects that the sampling points supply the minimum deviation of the total mean value of the outcome response with respect to the actual total mean value.

(C) **Robustness**

UED can be used to a variety of condition and is robust with respect to changes of model.

7.2.2 *Fundamental Principle of Uniform Experimental Design*

The fundamental principle of uniform experimental design includes the following terms:

(1) Total Mean Model

The basic assumption is that there exists a decisive relationship between the independent input variables $x_1, x_2, x_3, \ldots, x_s$ and the response f, which is expressed by a function,

$$f = F(x_1, x_2, x_3, \ldots, x_p), \quad X = \left\{x_1, x_2, x_3, \ldots, x_p\right\} \in C^p \tag{7.3}$$

Furthermore, it assumes that the domain of the independent input variables is in the unit cube $C^p = [0, 1]^p$, and thus, the total mean value of the response y on C^p is,

$$\overline{E(f)} = \int_{C^p} F(x_1, x_2, x_3, \ldots, x_p) \cdot \mathrm{d}x_1 \cdot \mathrm{d}x_1 \cdot \mathrm{d}x_3 \cdots \mathrm{d}x_p \tag{7.4}$$

While, if n sampling points $q_1, q_2, p_3, \ldots, q_n$ are taken on C^p, then the mean value of f on these n sampling points is,

$$\overline{f(D_n)} = \frac{1}{n}\sum_{j=1}^{n} F(q_j) \tag{7.5}$$

In Eq. (7.5), $D_n = \{q_1, q_2, q_3, \ldots, q_n\}$ indicates a design of these n sampling points.

Fang and Wang further showed that if the distribution of the sampling points $q_1, q_2, q_3, \ldots, q_n$ is uniform within the domain C^p, the deviation $\overline{E(f)} - \overline{f(D_n)}$ of the sampling point set on C^p and D_n is very small.

(2) **Uniform Design Table**

A series of "uniform design table" and their utility tables were developed by Fang and Wang for the appropriate application of UED, which can be employed to determine the position of sampling points conveniently.

However, it is not determined that how many sampling points are needed for complicated integration directly, here in this Chapter, the necessary number of sampling points with certain accuracy for estimation of definite integral is studied from the viewpoint of practical application.

(3) **Regression**

Finally, approximate expression for response $f' = F'(x_1, x_2, x_3, \ldots, x_p)$ with respect to the independent input variables of the sampling points could be fitted to reflect the resemble formation.

Surely, the approximate expression for response of the total preferable probability with respect to the independent input variables of the sampling points could be fitted to gain its resemble formation afterward as well.

7.3 Feature Analysis of the Periodic Function in a Single Period

(1) **One variable (1-D) case**

In general, the value of a function (e.g., integrand) evolutes from point to point according to a certain regulation within a domain. An example is the sine function, which is represented as,

$$f(x) = B \cdot [1 + \mathrm{Sin}(2\pi x/C)] \tag{7.6}$$

In Eq. (7.6), B expresses the amplitude coefficient, C represents the period (wave length) of this sine function, and x reflects the coordinate value in 1-D.

Obviously, the peak value of this function is $f = 2B$ at $x = x_0 = C/4$. While at another position $x_1 = x_0 + \Delta x/2$, $\Delta x/2$ is the distance from x_0, the value of the function f changes to, $f_1 = B \cdot [1 + \mathrm{Sin}(2\pi x_1/C)] = B \cdot \{1 + \mathrm{Sin}[2\pi(x_0 + \Delta x/2)/C]\} = B \cdot [1 + \mathrm{Sin}(\pi/2 + \Delta x\pi/C)]$.

If $\Delta x\pi/C = 0.2856$ rad, the function f takes the value of $f_1 = 1.96B$, which results in less error or deviation with respect to its peak value of $2B$.

Above analysis shows that if one wants to provide an approximation value of the periodic function $f(x)$ with less error or deviation with respect to its peak value by sub-cutting the period, the partition number κ_1 of the sub-cut in the period range (wave length) C for this periodic function within its single period is,

$$\kappa_1 = C/\Delta x = \pi/0.2856 \cong 11. \tag{7.7}$$

At the same time, the distance between the nearest neighbor sampling points is $\Delta x = C/11$.

Equation (7.7) shows that the 11 sampling points in one single periodic range (wave length) could supply an efficient approximation to the peak value with less error or deviation to its peak value of the function in one dimension.

(2) **Two variables (2-D) problem**

In the case of 2-D, the problem is on a plane where a rectangular coordinate system could be established, which consists of two orthogonal coordinate axes, says X- and Y-axes.

First, lets' employ the preliminary condition in the uniform experimental design method only [2, 3], which states that the projections of any two sample points on one coordinate axis are not coincide. We are now discussing the worst case, which is the status that all the sampling points are distributed along the diagonal line of the square; see Fig. 7.1.

Even in this worst case, the distance between the nearest neighbor sampling points is enlarged by $\sqrt{1^2 + 1^2} = \sqrt{2}$ times as compared to that of the distance between the nearest neighbor sampling points of 1-D case. So, if one tries to supply an appropriate approximation with a similar precision as that in one-dimensional case for the function, the sub-cut should be refined by about $1/\sqrt{2}$ times, let us take a factor 1/1.5, which results in the number of sampling points κ_2 to the period (wave length) C range of this periodical function within one period to be

$$\kappa_2 = 1.5 \times \kappa_1 = 1.5 \times 11 = 16.5 \cong 17. \tag{7.8}$$

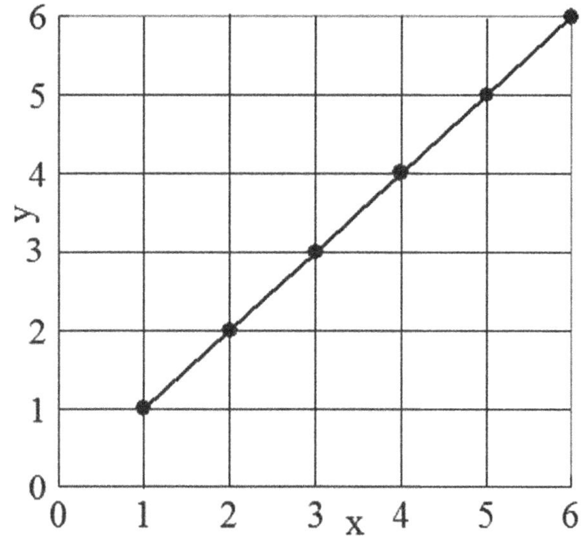

Fig. 7.1 Distribution of sampling points along the diagonal line of the square in 2-D

Equation (7.8) shows that 17 sampling points for two dimensions in one single periodic range (wave length) could supply an appropriate approximation for the peak value of the sine function with a similar precision like that in 1-D to its peak value for the sampling points along the diagonal line of the square roughly.

Next, one could employ the second demand of uniform experimental design that the sampling points must meet spatial filling or spatial uniformity besides projection properties. Thus, one could adjust the spatial distributions of the sampling points so that the distributions of the sampling points meet the demand of spatial uniformity simultaneously [2, 3].

Recalling Ripley's discussion that the expectation of the mean square error of the samples decreases with the spatial correlation in the problem of spatial sampling [12], it may result in the situation that the necessary number of sampling decreases with the spatial correlation of the samples. This may be related to the counterintuitive phenomena of using QMC in high dimensions with certain number of sampling points, since the sampling points may be highly correlated in high dimensions in problems of the previous section.

(3) **Three-dimensional problem**

Analogically, in three-dimensional problem, i.e., cube, a rectangular coordinate system is established, which consists of three orthogonal coordinate axes, i.e., *X*-, *Y*-, and *Z*-axes in general. Again, let us begin our discussion from the worst case. As the distribution of all the sampling points is conducted along the diagonal line of the cube; see Fig. 7.2.

In this worst condition, the distance between the nearest neighbor sampling points is enlarged by $\sqrt{1^2 + 1^2 + 1^2} = \sqrt{3}$ times as compared to that of the distance between the nearest neighbor sampling points of one dimension. Therefore, if one

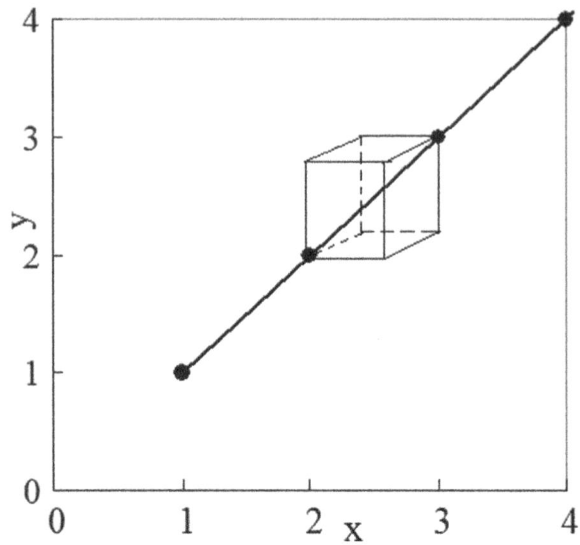

Fig. 7.2 Distribution of all the sampling points along the diagonal line of the cube in 3-D

attempts to supply an appropriate approximation for the peak value of the function with a similar precision as that of the one-dimensional case for the function once more, the sub-cut is refined by about 1/1.732 times, which leads to the number of sampling points κ_3 to the period (wave length) C range of this periodical function within one period

$$\kappa_3 = 1.732 \times \kappa_1 = 1.732 \times 11 = 19.052 \cong 19. \tag{7.9}$$

Equation (7.9) shows that the 19 sampling points of the one single periodic range (wave length) could supply an accurate approximation for the peak value of the sine function with a similar precision like that in 1-D to its peak value in three dimensions for the sampling points along the diagonal line of the cube assumedly.

Afterward, one could adjust the spatial distributions of the sampling points according to the second demand of the uniform design method [2, 3].

From the above discussion, it shows that if one attempts to provide an appropriate estimation for a periodical function in its single peak domain, 11 sampling points in 1-D case, 17 sampling points in 2-D case, and 19 sampling points in 3-D case are assumedly quite necessary for the estimation of the peak value of the function, respectively; and the sampling points must be deterministically distributed according to the rule of GLP and uniform design method. In the following sections, we will check the applicability of the above descriptions by typical examples in approximation for definite integral.

7.4 Typical Examples for the Efficient Approach of Numerical Integration for a Single Peak Function Based on Rules of GLD and Uniform Design Method

According to Hua and Wang, a set of good lattice points (GLP) with low discrepancy could provide an efficient value for a definite integral [2–4], and the discrepancy of the summation of its function values in the discretized GLPs with respect to the precise value of definite integral in one dimension is not greater than $\eta = D(m) \cdot V(f)$, where $D(m)$ is the discrepancy of the m sampling points of the point set, and $D(m) = O(m^{-1})$, $V(f)$ is the variation of the function $f(x)$ in its domain by m sampling points which are uniformly distributed [2–4].

The discussions in the previous sections show that 11 sampling points of the circumference in the 1-D case could supply an appropriate estimation for the peak value of the function with less error with respect to its real peak value provided they are uniformly distributed. Therefore, the error of the summation of the discretized sinusoidal function in the GLPs with respect to its precise value of integration is expected very small in 1-D case as well, more discussion will be found in Chap. 13.

Analogically, it supposes that 17 and 19 uniformly distributed sampling points in one single periodic range could supply an appropriate estimation with a similar precision like that in 1-D to its peak value in 2-D and 3-D cases, respectively, and furthermore an appropriate estimation for definite integral can be conducted correspondingly.

However, in general, other functions could be expanded as sine or cosine functions.

Hence, in the following sections, we will conduct some typical examples of definite integrals to present the applicability of the approach in viewpoint of practical application. The sampling points follow the rules of GLP so as to give low discrepancy [2–4].

(1) **1-D cases**

(A1) **Estimation for the probability integral**

The first example is the estimation of probability integral [16],

$$\int_0^{\infty} \exp(-x^2) \cdot \mathrm{d}x = \frac{\sqrt{\pi}}{2} \approx 0.886227 \tag{7.10}$$

i.e.,

$$\int_0^{\infty} y(x) \cdot \mathrm{d}x \equiv \int_0^{\infty} \exp(-x^2) \cdot \mathrm{d}x \approx 0.886227 \tag{7.11}$$

In Eq. (7.11), the integrand function is $y(x) = \exp(-x^2)$. As to $y(x) = \exp(-x^2)$, at $x_u = 4$ it gets a very small value of $y(x_u) = 1.125 \times 10^{-7}$, and therefore the upper limit of the integral could be set as $x_u = 4$ with very good accuracy.

Furthermore, according to the uniform experimental design method [2, 3], the position of the sampling points within the integral domain of [0, 4] is determined, which is shown in Table 7.1, and the integration Eq. (7.11) is thus discretized as,

$$I_0 = \int_0^4 y(x) \cdot \mathrm{d}x \approx \frac{4}{11} \sum_{i=1}^{11} y(x_i) \tag{7.12}$$

The position of the sampling points within the domain [0, 4] is determined by the following rule according to the principle of uniform distribution [2–4],

$$x_i = 4 \times (2i - 1)/(2 \times 11), \quad i \in 1, 2, 3, \ldots, 11. \tag{7.13}$$

The summated result of the right-hand side of Eq. (7.12) gives a value of 0.886227, which exactly equals to the real value of 0.886227 by chance [12], and indicates a higher accuracy.

(A2) **Estimation of the elliptic integral calculus of the magnetic field intensity for an elliptical current-carrying ring**

The next example is the estimation of an integral for an elliptical current-carrying ring, which is with the half major axis a, the half minor axis b, and the distance c is from the center O to the focal point F; the distance of a point Q to the center O of the ellipse is r, as shown in Fig. 7.3.

The problem is to get the magnetic field intensity at the center point O of the elliptical current-carrying ring.

Solution

In polar coordinate system, the equation of the ellipse with the center point 0 is

$$r = \sqrt{a^2 \cos^2 \phi + b^2 \sin^2 \phi} = a\sqrt{1 - \chi^2 \sin^2 \phi}, \tag{7.14}$$

in Eq. (7.14), $\chi \equiv c/a = (a^2 - b^2)^{0.5}/a$.

Table 7.1 Locations of the sampling points within the integral domain [0, 4]

Point No.	1	2	3	4	5
Location	0.182	0.545	0.909	1.273	1.636

Point No.	6	7	8	9	10	11
Location	2.0	2.364	2.727	3.091	3.455	3.818

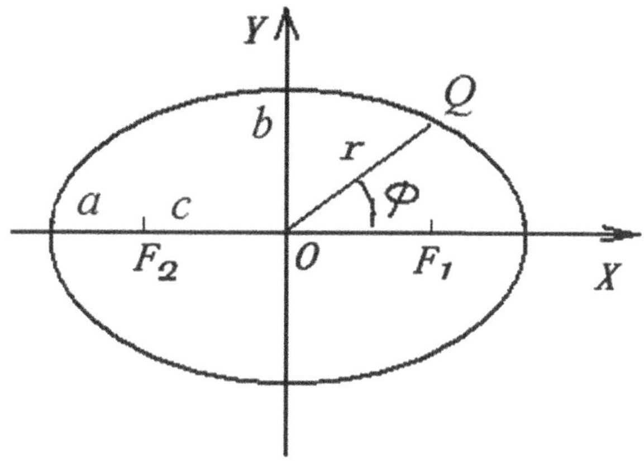

Fig. 7.3 Polar coordinate of the elliptical current-carrying ring

Thus, the magnetic field intensity at the center point O of the current-carrying ellipse can be expressed as [17],

$$B = \frac{\mu_0 I}{4\pi a}\int_0^{2\pi}\frac{\mathrm{d}\phi}{\sqrt{1-\chi^2\sin^2\phi}} = \frac{\mu_0 I}{\pi a}\int_0^{\pi/2}\frac{\mathrm{d}\phi}{\sqrt{1-\chi^2\sin^2\phi}} \tag{7.15}$$

In Eq. (7.15), μ_0 and I represent the intensity of the permeability of vacuum and the electric current, respectively.

Let Q represent the integration part in Eq. (7.15), i.e., $Q = \int_0^{\pi/2}\frac{\mathrm{d}\phi}{\sqrt{1-\chi^2\sin^2\phi}} = \int_0^{\pi/2} q(\phi)\cdot \mathrm{d}\phi$ with $q(\phi) = \frac{1}{\sqrt{1-\chi^2\sin^2\phi}}$, then Eq. (7.15) can be rewritten as

$$B = \frac{\mu_0 I}{\pi a}Q \tag{7.16}$$

In condition of $\chi = 0.3$, one could estimate the value of Q by our approximate approach.

Again, in light of uniform experimental design method [2, 3], the distribution of the sampling points in the integral domain $[0, \pi/2]$ can be obtained and shown in Table 7.2, and thus the integration Eq. (7.16) is discretized as,

$$Q = \int_0^{\pi/2} q(\phi)\cdot \mathrm{d}\phi \approx \frac{\pi/2}{11}\sum_{i=1}^{11} q(\phi_i) \tag{7.17}$$

The summation of the right-hand side of Eq. (7.17) gains a value of 1.608049, which equals to the exact value of the elliptic integral of 1.608049 luckily [17, 18], and shows a much higher accuracy of this approximate approach.

Table 7.2 Distribution of the sampling points in the integral domain $[0, \pi/2]$

Point No.	1	2	3	4	5	
Location	0.0714	0.2142	0.3570	0.4998	0.6426	
Point No.	6	7	8	9	10	11
Location	0.7854	0.9282	1.0710	1.2138	1.3566	1.4994

(2) **2-D problems**

In condition of 2-D or 3-D cases, a series of uniform design tables and their utility tables were developed by Fang and Wang in accordance with GLP and number-theoretic methods [2, 3], which are specific tables for uniform experimental design. The uniform design table $U^*_{17}(17^5)$ is the appropriate selection here for our usage, which includes 17 sampling points.

Now take the integration of $\int_{x_1=1.4}^{2.0} \mathrm{d}x_1 \int_{x_2=1.0}^{1.5} \ln(x_1 + 2x_2) \cdot \mathrm{d}x_2$ as our example.

Let J represents the integration, i.e., $J = \int_{x_1=1.4}^{2.0} dx_1 \int_{x_2=1.0}^{1.5} J(x_1, x_2)\cdot dx_2 \equiv \int_{x_1=1.4}^{2.0} dx_1 \int_{x_2=1.0}^{1.5} \ln(x_1 + 2x_2)\cdot dx_2$. The integration is with the precise value of 0.429560 [19].

The distribution of the sampling points is shown in the integral domain $[1.4, 2.0] \times [1.0, 1.5]$ in Table 7.3. The symbols x_{10} and x_{20} in Table 7.3 show the original positions from the uniform design table $U^*_{17}(17^5)$ in domain of $[1, 17] \times [1, 17]$ [2, 3].

In accordance with the uniform experimental design method [2, 3], the integration J in the domain $[1.4, 2.0] \times [1.0, 1.5]$ is discretized as

$$J \approx \frac{0.6 \times 0.5}{17} \sum_{j=1}^{17} J(x_{1j}, x_{2j}) \tag{7.18}$$

The summated result of the right-hand side of Eq. (7.18) gives a value of 0.429609, which has a relative error of $1.14 \times 10^{-4}\%$ with respect to its actual value of 0.429560 [19].

(3) **3-D cases**

(A) Integration $S = \int_{x_1=0}^{1} \mathrm{d}x_1 \int_{x_2=0}^{1} \mathrm{d}x_2 \int_{x_3=0}^{1} (x_1^3 + x_1 \cdot x_2^3 \cdot x_3^2 + x_3) \cdot \mathrm{d}x_3$.

The integration $S = \int_{x_1=0}^{1} \mathrm{d}x_1 \int_{x_2=0}^{1} \mathrm{d}x_2 \int_{x_3=0}^{1} (x_1^3 + x_1 \cdot x_2^3 \cdot x_3^2 + x_3) \cdot \mathrm{d}x_3$ was conducted by Chen et al. to study the effectiveness of the integration of multivariate functions with orthogonal arrays [20]. Now take it as an example by using our newly developed approximate approach for the definite integral.

The integration of $S = \int_{x_1=0}^{1} \mathrm{d}x_1 \int_{x_2=0}^{1} \mathrm{d}x_2 \int_{x_3=0}^{1} S(x_1, x_2, x_3) \cdot \mathrm{d}x_3 \equiv \int_{x_1=0}^{1} \mathrm{d}x_1 \int_{x_2=0}^{1} \mathrm{d}x_2 \int_{x_3=0}^{1} (x_1^3 + x_1 \cdot x_2^3 \cdot x_3^2 + x_3) \cdot \mathrm{d}x_3$ has the precise value of 19/24 = 0.791667 [20].

The uniform design table $U^*_{19}(19^7)$ is the proper selection for this usage, which includes 19 partition points. The position of the sampling points in the integral domain

Table 7.3 Position of the sampling points in the integral domain [1.4, 2.0] × [1.0, 1.5]

No.	x_{10}	x_{20}	x_1	x_2
1	1	7	1.4176	1.1912
2	2	14	1.4529	1.3971
3	3	3	1.4882	1.0735
4	4	10	1.5235	1.2794
5	5	17	1.5588	1.4853
6	6	6	1.5941	1.1618
7	7	13	1.6294	1.3676
8	8	2	1.6647	1.0441
9	9	9	1.7000	1.2500
10	10	16	1.7353	1.4559
11	11	5	1.7706	1.1324
12	12	12	1.8059	1.3382
13	13	1	1.8412	1.0147
14	14	8	1.8765	1.2206
15	15	15	1.9118	1.4265
16	16	4	1.9471	1.1029
17	17	11	1.9824	1.3088

[0, 1] × [0, 1] × [0, 1] is conducted and shown in Table 7.4, in which x_{10}, x_{20}, and x_{30} represent the original locations in the uniform design table $U^*_{19}(19^7)$ in the domain of [1, 19] × [1, 19] × [1, 19] [2, 3].

Following the uniform experimental design method, the integration S in the integral domain [0, 1] × [0, 1] × [0, 1] can be discretized as,

$$S \approx \frac{1}{19}\sum_{j=1}^{19} S(x_{1j}, x_{2j}, x_{3j}) \tag{7.19}$$

The summated result of the right-hand side of Eq. (7.19) leads to a value of 0.801534, which gains a relative error of 1.25% with respect to its actual value of 0.791667, while Chen et al. obtained a relative error of 0.04% by simulation calculation with 100 tests in $L_{100}(2^{99})$ orthogonal arrays [20]. Clearly, their simulation calculation amount is inevitably huge.

(B) **Integration** $K = (\frac{1}{2\pi})^{3/2} \cdot \int_{x_1=0}^{1} dx_1 \int_{x_2=0}^{1} dx_2 \int_{x_3=0}^{1} \exp[-\frac{1}{2}(x_1^2 + x_2^2 + x_3^2)] \cdot dx_3$.

The integration $K = \int_{x_1=0}^{1} dx_1 \int_{x_2=0}^{1} dx_2 \int_{x_3=0}^{1} K(x_1, x_2, x_3) \cdot dx_3 \equiv (\frac{1}{2\pi})^{3/2} \cdot \int_{x_1=0}^{1} dx_1 \int_{x_2=0}^{1} dx_2 \int_{x_3=0}^{1} \exp[-\frac{1}{2}(x_1^2 + x_2^2 + x_3^2)] \cdot dx_3$ was taken by Zheng et al. as an example to study the effectiveness of his algorithm with Monte Carlo [21], and Han et al. comparatively conducted the error analysis of quasi-Monte Carlo and

Table 7.4 Position of the sampling points in the integral domain [0, 1] × [0, 1] × [0, 1]

No.	x_{10}	x_{20}	x_{30}	x_1	x_2	x_3
1	1	11	13	0.0263	0.5526	0.6579
2	2	2	6	0.0789	0.0789	0.2895
3	3	13	19	0.1316	0.6579	0.9737
4	4	4	12	0.1842	0.1842	0.6053
5	5	15	5	0.2368	0.7636	0.2368
6	6	6	18	0.2895	0.2895	0.9211
7	7	17	11	0.3421	0.8684	0.5526
8	8	8	4	0.3947	0.3947	0.1842
9	9	19	17	0.4474	0.9737	0.8684
10	10	10	10	0.5000	0.5000	0.5000
11	11	1	3	0.5526	0.0263	0.1316
12	12	12	16	0.6053	0.6053	0.8158
13	13	3	9	0.6579	0.1316	0.4474
14	14	14	2	0.7105	0.7105	0.0789
15	15	5	15	0.7632	0.2368	0.7632
16	16	16	8	0.8158	0.8158	0.3947
17	17	7	1	0.8684	0.3421	0.0263
18	18	18	14	0.9211	0.9211	0.7105
19	19	9	7	0.9737	0.4474	0.3421

Monte Carlo integrals with this function [22]. Here the reanalysis to this integration is conducted by using our newly developed appropriate algorithm.

The explicit value of this integration of $K = (\frac{1}{2\pi})^{3/2} \cdot \int_{x_1=0}^{1} \mathrm{d}x_1 \int_{x_2=0}^{1} \mathrm{d}x_2 \int_{x_3=0}^{1} \exp[-\frac{1}{2}(x_1^2 + x_2^2 + x_3^2)] \cdot \mathrm{d}x_3$ is 0.039772 [21]. The uniform experimental design table $U^*_{19}(19^7)$ again is the proper selection for this three-dimensional problem. The positions of sampling points in the integral domain [0, 1] × [0, 1] × [0, 1] are presented in Table 7.4 again [2, 3].

Once more, according to uniform experimental design [2, 3], the integration K in the integral domain [0, 1] × [0, 1] × [0, 1] is discretized as,

$$K \approx \frac{1}{19} \sum_{j=1}^{19} K(x_{1j}, x_{2j}, x_{3j}) \tag{7.20}$$

The summated result of right hand of Eq. (7.20) gains a value of 0.039852, which leads to a relative error of 0.20% with respect to its actual value of 0.039772, while Zheng et al. obtained a result of 0.039772 with 250 samplings with Monte Carlo simulations [21], Han et al. gained relative error of 0.16% by employing Monte Carlo simulations with 597 samplings and 0.14% by employing quasi-Monte Carlo simulations with 101 samplings [22]. Explicitly, their simulation calculation amounts are extremely huge.

(4) **Summary**

The efficient approach to estimate a definite integral with finite sampling points shows very good behavior in viewpoint of practical application. An approximate result for a definite integral can be obtained within its single peak domain by using the new approach with 11 sampling points for 1-D, 17 sampling points for 2-D, and 19 sampling points for 3-D cases assumedly. The distribution of sampling points is deterministic and uniform in accordance with the rule of the "good lattice points" and uniform experimental design method. The efficient approach is beneficial to relevant application and research [23].

7.5 Typical Examples of Applications of the Finite Sampling Point Method in Assessment of Probability-Based Multi-objective Optimization

The above statements illuminate the remarkable features of uniform experiment design, i.e., uniform distribution of sampling/experiment points in the test domain and the small number of tests, sufficient representative of each point, and easy to conduct regression analysis, etc., so the finite sampling point method can be helpful to simplify the complicated data process in evaluation of material selection with probability-based multi-objective optimization. In the following section, some examples are given.

(1) **Simplified assessment for multi-objective optimization with single variable**

The multi-objective optimization with single variable is relatively simple, which corresponds to 1-D problem.

The example here is that the simultaneous optimization of both min $y_1(x) = x^2$ and min $y_2(x) = (x - 2)^2$ in the domain of $x \in [-5, 7]$, which was analyzed by Huang et al. with complex and tediously long evolutionary computations by using Pareto optimization [24].

Here, in the respect of probability-based approach for multi-objective optimization, the partial preferable probabilities for $y_1(x)$ and $y_2(x)$ can be expressed as,

$$P_{f1} = \left(49 - x^2\right)/432, \quad P_{f2} = \left[49 - (x - 2)^2\right]/432. \tag{7.21}$$

Moreover, the total preferable probability $P_t = P_{f1} \bullet P_{f2}$ takes its maximum value at $x = 1$ clearly. Correspondingly, the simultaneous minimum values of $y_1(x)$ and $y_2(x)$ are 1 compromisingly at $x = 1$, respectively. Obviously, the evaluating process is much easier than that of complex evolutionary computations by using Pareto optimization [24].

Table 7.5 Positions of the distribution of the sampling points within the integral domain [− 5, 7] together with the value of P_t and the ranking

No.	Position of point	$P_t \times 10^2$	Rank
1	− 4.4546	0.1147	6
2	− 3.3636	0.4085	5
3	− 2.2727	0.7221	4
4	− 1.1818	0.9916	3
5	− 0.0909	1.1716	2
6	1.0000	1.2346	1
7	2.0909	1.1716	2
8	3.1818	0.9916	3
9	4.2727	0.7221	4
10	5.3636	0.4085	5
11	6.4546	0.1147	6

Meanwhile, if the sampling point method is employed [23], 11 sampling points can be used to deal with this problem. The uniform distribution of the sampling points is shown in Table 7.5 within the domain of $x \in [-5, 7]$ together with the value of the total preferable probability P_t and the consequence of ranking.

Obviously, From Table 7.5 P_t gains its maximum value at $x = 1$ exactly.

(2) **Simplified assessment of multi-objective optimization for tower crane boom tie rods**

Qu et al. once performed the assessment of multi-objective optimization for tower crane boom tie rods by fuzzy optimization model [25].

With their careful analysis, optimal objectives are the minimum mass $M(X)$ of the boom tie rod together with the minimum angular displacement $\psi(X)$ of the boom. The models for $M(X)$ and $\psi(X)$ are expressed as,

$$M(X) = 208.323x_1 + 433.868x_2 \tag{7.22}$$

$$\psi(X) = \frac{2.0288 \times 10^{-4}}{9.8621x_1 + 5.3471x_2} \tag{7.23}$$

The domain of the variables is,

$$0.003379 < x_1 < 0.005805, \tag{7.24}$$

$$0.003379 < x_2 < 0.005468. \tag{7.25}$$

According to Qu's optimal requirements for $M(X)$ and $\psi(X)$ [25], both $M(X)$ and $\psi(X)$ belong to unbeneficial type of indexes, which have the characteristic of the smaller the better in the optimal process.

Therefore, in accordance with the probability-based multi-objective optimization, the partial preferable probabilities of both $M(X)$ and $\psi(X)$ can be formulated as

$$P_M = \beta_M \cdot [M_{max} + M_{min} - M(X)], \tag{7.26}$$

$$P_\psi = \beta_\psi \cdot [\psi_{max} + \psi_{min} - \psi(X)]. \tag{7.27}$$

In Eqs. (7.26) and (7.27), M_{max}, M_{min}, and β_M express the maximum and minimum values of the index $M(X)$, and the normalization factor, respectively; ψ_{max}, ψ_{min}, and β_ψ indicate maximum and the minimum values of the index $\psi(X)$, and the normalization factor, individually.

Meanwhile,

$$\beta_M = \frac{1}{\int_{x_{1l},x_{2l}}^{x_{1u},x_{2u}} [M_{max} + M_{min} - M(X)] \cdot dx_1 \cdot dx_2} \tag{7.28}$$

$$\beta_\psi = \frac{1}{\int_{x_{1l},x_{2l}}^{x_{1u},x_{2u}} [\psi_{max} + \psi_{min} - \psi(X)] \cdot dx_1 \cdot dx_2} \tag{7.29}$$

In Eqs. (7.28) and (7.29), x_{1L}, x_{1U}, x_{2L}, and x_{2U} indicate the lower limit and upper limit of variables x_1 and x_2 in their domains, respectively.

According to the common procedure, the next thing is to substitute Eqs. (7.22) and (7.23) into Eqs. (7.26) through (7.29) to perform the evaluations with the constraints of Eqs. (7.24) and (7.25). It can be seen that the evaluations are complex and tediously long. However, the finite sampling point algorithm can be used, and the definite integral in Eqs. (7.28) and (7.29) can thus be simplified with discrete sampling points.

According to finite sampling point algorithm [23], 17 discrete sampling points are necessary for the two independent variables x_1 and x_2. Therefore, the uniform design table of $U_{17}^*(17^5)$ is employed to perform the assessment. Table 7.6 shows the designed results together with calculated data for $M(X)$ and $\psi(X)$, while the symbols x_{10} and x_{20} represent the positions of sampling points in the original domain of $[1, 17] \times [1, 17]$ of uniform design table $U_{17}^*(17^5)$.Table 7.7 presents the assessments of this problem.

Table 7.7 indicates that the evaluated result of *No. 13* exhibits the maximum value of total preferable probability at first glance; therefore, the optimal configuration could be around test *No. 13*.

As to the sampling point *No. 13* in Table 7.6, which is at $x_1 = 0.0052$ m^2 and $x_2 = 0.0034$ m^2, the corresponding mass $M_{optim.}$ of the boom tie rod and the corresponding angular displacement $\psi_{optim.}$ of the boom are 2.5682 tons and 0.0029°, respectively, which are superior to those of Qu's consequences of 2.8580 tons, and 0.0026° at $x_1 = 0.0058$ m^2 and $x_2 = 0.0038$ m^2, comprehensively.

Furthermore, regression analysis can be conducted to perform further optimization. The regressed consequence of the total probability P_t can be expressed

Table 7.6 Designed results $U_{17}^{*}(17^5)$ together with calculated data for $M(X)$ and $\psi(X)$

No.	x_{10}	x_{20}	x_1/m^2	x_2/m^2	M/t	Ψ/°
1	1	7	0.0035	0.0042	2.5314	0.0036
2	2	14	0.0036	0.0050	2.9343	0.0033
3	3	3	0.0037	0.0037	2.3776	0.0036
4	4	10	0.0039	0.0045	2.7805	0.0032
5	5	17	0.0040	0.0054	3.1834	0.0030
6	6	6	0.0042	0.0041	2.6267	0.0032
7	7	13	0.0043	0.0049	3.0296	0.0030
8	8	2	0.0044	0.0036	2.4729	0.0032
9	9	9	0.0046	0.0044	2.8758	0.0029
10	10	16	0.0047	0.0053	3.2788	0.0027
11	11	5	0.0049	0.0039	2.7220	0.0029
12	12	12	0.0050	0.0048	3.1250	0.0027
13	13	1	0.0052	0.0034	2.5682	0.0029
14	14	8	0.0053	0.0043	2.9712	0.0027
15	15	15	0.0054	0.0052	3.3741	0.0025
16	16	4	0.0056	0.0038	2.8174	0.0027
17	17	11	0.0057	0.0047	3.2203	0.0025

as,

$$P_t \times 10^3 = 8.2971 - 249.4110x_1 - 304.5570x_2 - 0.0978 \times 10^{-1}x_1^{-1} - 0.0083 \times 10^{-1}x_2^{-1}, \tag{7.30}$$

$$R^2 = 0.9362. \tag{7.31}$$

At the same time, the regressed consequence of $M(X)$ is expressed as,

$$M(X) = 2.89 \times 10^{-15} + 208.3230x_1 + 433.8680x_2, \tag{7.32}$$

$$R^2 = 1. \tag{7.33}$$

The regressed consequence of the ψ with respect to x_1 and x_2 is expressed as,

$$\psi(X) = 0.0035 - 0.1459x_1 - 0.2412x_2 - 5.7700 \times 10^{-6}x_1^{-1} - 1.4000 \times 10^{-7}x_2^{-1}, \tag{7.34}$$

$$R^2 = 0.9941. \tag{7.35}$$

Table 7.7 Assessments of this problem

No.	Partial preferable probability		Total	
	$P_{M(x)}$	$P_{\psi(X)}$	$P_t \times 10^3$	Rank
1	0.0659	0.0471	3.1006	16
2	0.0576	0.0536	3.0905	17
3	0.0690	0.0473	3.2642	13
4	0.0608	0.0538	3.2703	12
5	0.0525	0.0592	3.1091	15
6	0.0639	0.0540	3.4512	8
7	0.0557	0.0593	3.3037	11
8	0.0671	0.0542	3.6333	5
9	0.0588	0.0595	3.4993	7
10	0.0506	0.0639	3.2346	14
11	0.0620	0.0596	3.6957	3
12	0.0537	0.0641	3.4426	9
13	**0.0651**	**0.0598**	**3.8930**	**1**
14	0.0569	0.0642	3.6514	4
15	0.0486	0.0680	3.3052	10
16	0.0600	0.0643	3.8609	2
17	0.0518	0.0681	3.5246	6

The optimal consequences of the regression formula Eq. (7.30) are $P_t^* \times 10^3 = 3.8890$, and correspondingly $M^* = 2.6754$ tons, $\psi^* = 0.0028°$ at $x_1 = 0.0058$ m^2 and $x_2 = 0.0034$ m^2, which are superior to those of Qu's consequences as well.

(3) **Optimal design for the composition of rubber**

Gao et al. raised a problem of optimal design for the composition of rubber with four desired conditions for the performance indexes [26], i.e., ultimate tensile strength $y_1 > 12$ MPa, tensile strain $y_2 > 600\%$, tear strength $y_3 > 40$ kN m^{-1}, and residual strain at fracture $y_4 < 34\%$. The ranges for chemicals as independent variables are, $20 \le x_1 \le 42$, $0.8 \le x_2 \le 2$, and $0.8 \le x_3 \le 2.2$ [26]. There exist four fitted relationships of the performance indexes *vs* independent variables, which are expressed as [26],

$$y_1 = 17.5 + 0.374x_1 - 3.66x_2 - 5.58x_3 + 0.06764x_1x_2 + 0.03164x_lx_3 + 0.08786x_2x_3 - 0.01036x_1^2 + 0.45x_2^2 + 1.35x_3^2, \quad (7.36)$$

$$y_2 = 607 + 17.0x_1 - 59.4x_2 - 11.2x_3 - 0.2766x_1x_2 + 0.1846x_1x_3 - 35.58x_2x_3 - 0.3071x_1^2 + 29.92x_2^2 - 1.46x_3^2, \quad (7.37)$$

$$y_3 = 55.4 + 1.91x_1 - 11.1x_2 - 20.6x_3 + 0.3689x_1x_2 + 0.3692x_1x_3 \\ - 3.5144x_2x_3 - 0.0517x_1^2 + 0.92x_2^2 + 1.69x_3^2, \quad (7.38)$$

$$y_4 = 27.0 + 0.43x_1 - 0.42x_2 - 10.7x_3 + 0.0307x_1x_2 - 0.132x_1x_3 \\ + 1.3179x_2x_3 + 0.0007x_1^2 + 0.19x_2^2 + 5.24x_3^2. \quad (7.40)$$

Now, lets' discretize this 3-D problem with the uniform design table $U^*_{19}(19^7)$ of uniform experimental design [2, 3], and 19 sampling points in its integral domain [20, 42] × [0.8, 2] × [0.8, 2.2]. Table 7.8 shows the distributions of the sampling points. Table 7.9 shows the results of responses at these sampling points.

Furthermore, by using the four desired conditions for the performance indexes, it obtains the four utilities of the corresponding performance indexes, as shown in Table 7.10. Thus, the partial preferable probability and the total preferable probability can be obtained, which are shown in Table 7.11.

Table 7.8 Distributions of the sampling points and values of the corresponding performance indexes

No.	x_{10}	x_{20}	x_{30}	x_1	x_2	x_3
1	1	11	13	20.5790	1.4632	1.7211
2	2	2	6	21.7368	0.8947	1.2053
3	3	13	19	22.8947	1.5895	2.1632
4	4	4	12	24.0526	1.0211	1.6474
5	5	15	5	25.2105	1.7158	1.1316
6	6	6	18	26.3684	1.1474	2.0895
7	7	17	11	27.5263	1.8421	1.5737
8	8	8	4	28.6842	1.2737	1.0579
9	9	19	17	29.8421	1.9684	2.0158
10	10	10	10	3.0000	1.4000	1.5000
11	11	1	3	32.1579	0.8316	0.9842
12	12	12	16	33.3158	1.5263	1.9421
13	13	3	9	34.4737	0.9579	1.4263
14	14	14	2	35.6316	1.6526	0.9105
15	15	5	15	36.7895	1.0842	1.8684
16	16	16	8	37.9474	1.7789	1.3526
17	17	7	1	39.1053	1.2105	0.8368
18	18	18	14	40.2632	1.9053	1.7947
19	19	9	7	41.4211	1.3368	1.2789

Table 7.9 Results of responses at these sampling points

No.	y_1	y_2	y_3	y_4
1	14.1911	682.7218	43.4258	33.3119
2	15.2950	743.3061	53.9784	30.6206
3	14.5287	654.1533	39.6830	39.2425
4	14.7682	720.9343	49.4840	33.8394
5	14.8027	732.0161	52.8759	30.9027
6	14.7866	691.8434	45.7045	39.2081
7	14.3471	697.9910	47.7548	34.0600
8	14.9297	744.8693	55.0478	32.3031
9	14.4368	657.2467	43.3484	39.3670
10	14.1549	710.3217	49.8898	34.8983
11	14.7790	760.4413	54.7990	34.2484
12	13.9252	669.0549	45.4466	39.6434
13	13.6850	725.3713	49.6042	36.2816
14	14.0337	727.5082	53.0431	33.6548
15	13.1360	683.5821	45.1241	40.4647
16	13.0109	680.7848	47.2215	35.6263
17	13.1137	717.1984	48.8710	35.9108
18	12.5333	627.3423	42.1147	39.7478
19	11.7717	669.9526	43.0126	37.3205

From Table 7.11, it can be seen that the configuration *Nos. 1, 2, 4, 5, 8,* and *14* are met the desired conditions for the performance indexes, so they are the optimal solutions of this problem.

(4) **Linear programming problem with domain in non-regular area**

A linear programming problem with the domain in non-regular area is expressed as follows [27]:

$$\max f_1 = 9x_1 + 10x_2 \tag{7.41}$$

$$\min f_2 = 4x_1 + 5x_2 \tag{7.42}$$

$$\text{s.t.} : (1) : x_1 + 4x_2 \leq 20; \tag{7.43}$$

$$(2) : 4x_1 + 4x_2 \leq 50; \quad x_1, \quad x_2 > 0. \tag{7.44}$$

The symbols “max” and “min” in Eqs. (7.41) and (7.42) represent that maximum and minimum of the functions are best for the optimization, respectively; s. t. in

Table 7.10 Utilities of the four performance indexes

No.	U_1	U_2	U_3	U_4
1	1	1	1	1
2	1	1	1	1
3	1	1	0	0
4	1	1	1	1
5	1	1	1	1
6	1	1	1	0
7	1	1	1	0
8	1	1	1	1
9	1	1	1	0
10	1	1	1	0
11	1	1	1	0
12	1	1	1	0
13	1	1	1	0
14	1	1	1	1
15	1	1	1	0
16	1	1	1	0
17	1	1	1	0
18	1	1	1	0
19	0	1	1	0

Eqs. (7.43) and (7.44) stand for "subject to," i.e., constraint condition. The domain for independent variables x_1 and x_2 is determined by s. t. 1 and 2, i.e., Eqs. (7.43) and (7.44), which are shown in Fig. 7.4 indicating the non-regular area *ABCD*. The corresponding virtual rectangular area is *ABED*. The ratio of the areas of the non-regular area *ABCD* to the corresponding virtual rectangular area *ABED* is 0.65. Therefore, if uniform sampling method is employed to conduct the evaluation of linear programming problem, 17 uniform sampling points are needed for this problem within domain of non-regular area *ABCD*, and the corresponding number of the sampling points 17/0.65 ≅ 26 is required for the corresponding virtual rectangular area *ABED* according to the general principle for uniform distribution of sampling points in uniform experimental design [2, 3]. Thus, the distribution of the 17 sampling points in the non-regular area *ABCD* can be approximately obtained by arranging the 26 sampling points within the virtual rectangular area *ABED* by using uniform design table $U_{26}{}^{*}(26^{11})$, which is shown in Fig. 7.4 and Table 7.12.

The discrete values of the objective functions $f_1(x_1, x_2)$ and $f_2(x_1, x_2)$ are shown in Table 7.13, together with the assessed consequences for their preferable probability.

From Table 7.13, it can be seen that the maximum value of total preferable probability P_t is at the discrete sampling point *No. 17*, so the optimal solution for this linear programming problem is at $x_1 = 7.9327$ and $x_2 = 0.2885$, and the optimal values of

Table 7.11 Partial preferable probability and total preferable probability

No.	P_1	P_2	P_3	P_4	$P_t \times 10^5$
1	**0.0556**	**0.0526**	**0.0556**	**0.1667**	**2.7106**
2	**0.0556**	**0.0526**	**0.0556**	**0.1667**	**2.7106**
3	0.0556	0.0526	0	0	0
4	**0.0556**	**0.0526**	**0.0556**	**0.1667**	**2.7106**
5	**0.0556**	**0.0526**	**0.0556**	**0.1667**	**2.7106**
6	0.0556	0.0526	0.0556	0	0
7	0.0556	0.0526	0.0556	0	0
8	**0.0556**	**0.0526**	**0.0556**	**0.1667**	**2.7106**
9	0.0556	0.0526	0.0556	0	0
10	0.0556	0.0526	0.0556	0	0
11	0.0556	0.0526	0.0556	0	0
12	0.0556	0.0526	0.0556	0	0
13	0.0556	0.0526	0.0556	0	0
14	**0.0556**	**0.0526**	**0.0556**	**0.1667**	**2.7106**
15	0.0556	0.0526	0.0556	0	0
16	0.0556	0.0526	0.0556	0	0
17	0.0556	0.0526	0.0556	0	0
18	0.0556	0.0526	0.0556	0	0
19	0	0.0526	0.0556	0	0

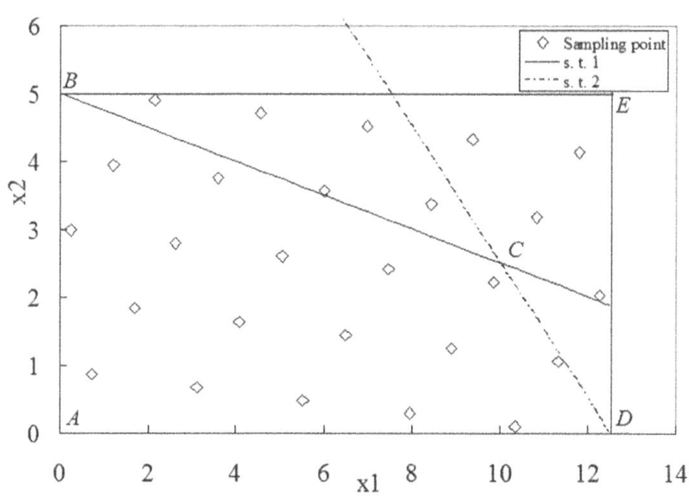

Fig. 7.4 Distribution of the sampling points in the linear programming problem

Table 7.12 Positions of the sampling points in the linear programming problem $U_{26}{}^{*}(26^{11})$ within domain [0, 12.5] × [0, 5]

No.	x_{10}	x_{20}	x_1	x_2	Notation
1	1	16	0.2404	2.9808	In ABCD
2	2	5	0.7212	0.8654	In ABCD
3	3	21	1.2019	3.9423	In ABCD
4	4	10	1.6827	1.8269	In ABCD
5	5	26	2.1635	4.9038	
6	6	15	2.6442	2.7885	In ABCD
7	7	4	3.1250	0.6731	In ABCD
8	8	20	3.6058	3.7500	In ABCD
9	9	9	4.0865	1.6346	In ABCD
10	10	25	4.5673	4.7115	
11	11	14	5.0481	2.5962	In ABCD
12	12	3	5.5288	0.4808	In ABCD
13	13	19	6.0096	3.5577	
14	14	8	6.4904	1.4423	In ABCD
15	15	24	6.9712	4.5192	
16	16	13	7.4519	2.4038	In ABCD
17	17	2	7.9327	0.2885	In ABCD
18	18	18	8.4135	3.3654	
19	19	7	8.8942	1.2500	In ABCD
20	20	23	9.3750	4.3269	
21	21	12	9.8558	2.2115	In ABCD
22	22	1	10.3365	0.0962	In ABCD
23	23	17	10.8173	3.1731	
24	24	6	11.2981	1.0577	In ABCD
25	25	22	11.7789	4.1346	
26	26	11	12.2596	2.0192	

objective function $f_1(x_1, x_2)$ and $f_2(x_1, x_2)$ are 74.2789 and 33.1731, individually. The results of the total preferable probability at the sampling points *Nos. 12* and *14* go after that of the sampling point *No. 17* tightly.

(5) Nonlinear programming problem with domain in non-regular area

A nonlinear programming problem with the domain in non-regular area is expressed as follows:

$$\max R(x_1, x_2) = 4x_1 + 2x_2 - 0.5x_1^2 - 0.25x_2^2; \qquad (7.45)$$

Table 7.13 Values of the objective functions $f_1(x_1, x_2)$ and $f_2(x_1, x_2)$, together with the assessed consequences for their preferable probability

No	Value of objective function		Preferable probability			Rank
	f_1	f_2	P_{f1}	P_{f2}	$P_t \times 10^3$	
1	31.9710	15.8654	0.0287	0.0892	2.5605	14
2	15.1442	7.2115	0.0136	0.1077	1.4638	17
3	50.2404	24.5192	0.0451	0.0708	3.1911	8
4	33.4135	15.8654	0.0300	0.0892	2.6760	13
6	51.6827	24.5192	0.0464	0.0708	3.2827	7
7	34.8558	15.8654	0.0313	0.0892	2.7915	11
8	69.9519	33.1731	0.0628	0.0523	3.2840	6
9	53.1250	24.5192	0.0477	0.0708	3.3743	4
11	71.3942	33.1731	0.0641	0.0523	3.3518	5
12	**54.5673**	**24.5192**	**0.0490**	**0.0708**	**3.4659**	**2**
14	**72.8365**	**33.1731**	**0.0654**	**0.0523**	**3.4195**	**3**
16	91.1058	41.8269	0.0818	0.0338	2.7676	12
17	**74.2789**	**33.1731**	**0.0667**	**0.0523**	**3.4872**	**1**
19	92.5481	41.8269	0.0831	0.0338	2.8114	10
21	110.8173	50.4808	0.0995	0.0154	1.5302	16
22	93.9904	41.8269	0.0844	0.0338	2.8552	9
24	112.2596	50.4808	0.1008	0.0154	1.5501	15

$$\max S(x_1, x_2) = 5.5 + 3x_1 + 1x_2 - 0.5x_1^2 - 0.25x_2^2; \tag{7.46}$$

$$\text{s. t. } 8x_1 + 5x_2 \leq 40; \quad x_1, x_2 > 0. \tag{7.47}$$

The domain for independent variables x_1 and x_2 is determined by s. t., i.e., Eq. (7.47), which is shown in Fig. 7.5 indicating the non-regular triangle area *ABD*. The corresponding virtual rectangular area is *ABED*. The ratio of the areas of the non-regular triangle area *ABD* to the corresponding virtual rectangular area *ABCD* is 0.5. Therefore, if uniform sampling method is employed to conduct the evaluation of nonlinear programming problem, 17 uniform sampling points are needed for this problem within domain of non-regular area *ABD*, and the corresponding number of the sampling points 17/0.5 = 34 is required for the corresponding virtual rectangular area *ABCD* according to the general principle for uniform distribution of sampling points in uniform experimental design [2, 3]. Thus, the distribution of the 17 sampling points in the non-regular triangle area *ABD* can be approximately obtained by arranging the 34 sampling points within the virtual rectangular area *ABCD*. Actually, the uniform experimental design method provides a quite near uniform table $U_{37}(37^{12})$, which is close to our demand and shown in Fig. 7.5 and Table 7.14.

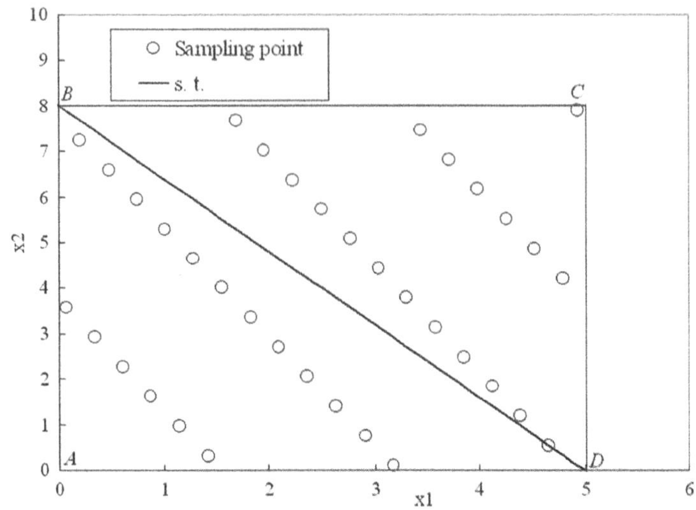

Fig. 7.5 Distribution of the sampling points in the nonlinear programming problem

The characteristic of this nonlinear programming problem is that the s. t. condition is linear and rather simple.

Shown in Table 7.15 are the discrete values of the objective functions $R(x_1, x_2)$ and $S(x_1, x_2)$, together with the assessed consequences for their preferable probability.

From Table 7.15, it can be seen that the sampling point *No. 18* exhibits the maximum value of total preferable probability P_t; therefore, the optimal solution for this nonlinear programming problem is at $x_1 = 2.3649$ and $x_2 = 2.0541$, and the optimal values of objective function $R(x_1, x_2)$ and $S(x_1, x_2)$ are 9.7165 and 10.7976, respectively. The results of the total preferable probability at sampling points *Nos. 16* and *20* follow that of the sampling point *No. 18* closely.

(6) Multi-objective optimization of numerical control machining parameters for high efficiency and low carbon

Li et al. conducted the multi-objective optimization of numerical control (NC) machining parameters for high efficiency and low carbon [28]. The optimal objectives contain the minimum processing time and lowest carbon emission, the latter includes carbon emissions of electric power, cutting tool carbon, and cutting fluid. This multi-objective optimization problem subjects to constraints from machine tool and processing property. Here, it is restudied by using the PMOO method with finite sampling point algorithm.

Features of Objectives

(A) Processing time

The total processing time includes cutting time, tool change time, and process auxiliary time. Maximum productivity (efficiency) can be achieved with the shortest

Table 7.14 Positions of the sampling points in the nonlinear programming problem in $U_{37}(37^{12})$ within domain [0, 5] × [0, 8]

No.	x_{10}	x_{20}	x_1	x_2	*Notation*
1	1	17	0.0676	3.5676	In ABD
2	2	34	0.2027	7.2432	In ABD
3	3	14	0.3378	2.9189	In ABD
4	4	31	0.4730	6.5946	In ABD
5	5	11	0.6081	2.2703	In ABD
6	6	28	0.7432	5.9459	In ABD
7	7	8	0.8784	1.6216	In ABD
8	8	25	1.0135	5.2973	In ABD
9	9	5	1.1486	0.9730	In ABD
10	10	22	1.2838	4.6486	In ABD
11	11	2	1.4189	0.3243	In ABD
12	12	19	1.5541	4.0000	In ABD
13	13	36	1.6892	7.6757	
14	14	16	1.8243	3.3514	In ABD
15	15	33	1.9595	7.0270	
16	16	13	2.0946	2.7027	In ABD
17	17	30	2.2297	6.3784	
18	18	10	2.3649	2.0541	In ABD
19	19	27	2.5000	5.7297	
20	20	7	2.6351	1.4054	In ABD
21	21	24	2.7703	5.0811	
22	22	4	2.9054	0.7568	In ABD
23	23	21	3.0405	4.4324	
24	24	1	3.1757	0.1081	In ABD
25	25	18	3.3108	3.7838	
26	26	35	3.4459	7.4595	
27	27	15	3.5811	3.1351	
28	28	32	3.7162	6.8108	
29	29	12	3.8514	2.4865	
30	30	29	3.9865	6.1622	
31	31	9	4.1216	1.8378	
32	32	26	4.2568	5.5135	
33	33	6	4.3919	1.1892	
34	34	23	4.5271	4.8649	
35	35	3	4.6622	0.5405	
36	36	20	4.7973	4.2162	
37	37	37	4.9324	7.8919	

Table 7.15 Values of the objective functions $P(x_1, x_2)$ and $Q(x_1, x_2)$, together with the assessed consequences for their preferable probability

No.	Value of objective function		Preferable probability			Rank
	$R(x_1, x_2)$	$S(x_1, x_2)$	P_R	P_S	$P_t \times 10^3$	
1	4.2212	6.0861	0.0344	0.0431	1.4818	16
2	2.1606	0.2147	0.0176	0.0015	0.0268	18
3	5.0021	7.2453	0.0407	0.0513	2.0904	14
4	4.0971	2.5295	0.0334	0.0179	0.5977	17
5	5.4995	8.1212	0.0448	0.0575	2.5761	13
6	5.7501	4.5609	0.0468	0.0323	1.5126	15
7	5.7136	8.7136	0.0465	0.0617	2.8715	10
8	7.1197	6.3089	0.0580	0.0447	2.5908	12
9	5.6442	9.0226	0.0460	0.0639	2.9373	9
10	8.2059	7.7735	0.0668	0.0550	3.6792	8
11	5.2914	9.0481	0.0431	0.0641	2.7615	11
12	9.0087	8.9546	0.0734	0.0634	4.6529	6
14	9.5280	9.8524	0.0776	0.0698	5.4145	4
16	**9.7640**	**10.4667**	**0.0795**	**0.0741**	**5.8945**	**2**
18	**9.7165**	**10.7976**	**0.0792**	**0.0764**	**6.0513**	**1**
20	**9.3856**	**10.8450**	**0.0765**	**0.0768**	**5.8709**	**3**
22	8.7713	10.6091	0.0715	0.0751	5.3673	5
24	7.8735	10.0898	0.0641	0.0714	4.5821	7

processing hours. The mathematical model of the processing time T_p can be expressed as [28],

$$T_p = \frac{\pi d_0 L_w \Delta}{1000 v_c f a_{sp}} + \frac{t_{ct} \pi d_0 L_w \Delta v_c^{x-1} f^{y-1} a_{sp}^{z-1}}{1000 C_T} + t_{ot} \tag{7.48}$$

In Eq. (7.48), t_{ct} is the time for changing tool, t_{oc} is other auxiliary time except tool change, L_w is the machining length, Δ is the machining allowance, d_0 is diameter of workpiece, v_c is the cutting speed, f is the feed amount, a_{sp} is the cutting depth, and C_t is the constant related to cutting conditions, x, y, z are coefficients for tool life.

(B) Carbon emissions

The carbon emissions in processing process mainly include the carbon emissions caused by the consumption of raw materials in the processing process, the carbon emissions caused by the consumption of electrical energy, the auxiliary materials used in the processing process (such as that from the use of tools and the use of cutting fluids), as well as the carbon emission caused by the post-processing of chips. The carbon emission function of the cutting process is [28],

$$
\begin{aligned}
C_p = 0.6747\{[P_{\text{uo}} + A_1\left(\frac{1000v_c}{\pi d_0}\right) + A_2\left(\frac{1000v_c}{\pi d_0}\right)^2]T_p \\
+ 1.2C_{\text{Fc}}a_{\text{sp}}^{x_{\text{FC}}} f^{y_{\text{Fc}}} K_{\text{Fc}} v_c^{(n_{\text{Fc}}+1)} t_m\} + 29.6\frac{t_m}{T_t} W_t \\
+ \frac{T_p}{T_t}\{2.85(C_c + A_c) + 0.2[\frac{(C_c + A_c)}{\delta}]\}
\end{aligned}
\tag{7.49}
$$

In Eq. (7.49), P_{uo} is the minimum non-loading power, A_1 and A_2 express coefficients of the spindle speed, W_t is the mass of tool, C_c represents the oil consumption of initial cutting, A_c represents the additional cutting oil consumption, δ is the concentration of cutting fluid, T_t is life of tool, t_m is cutting time, T_c is the duration of cutting fluid, F_t is a factor of carbon emission of tool (29.6 kg CO_2/kg), and T_p is processing time.

Constraint Conditions

Conclusively, the constraint conditions for the high efficiency and low carbon optimization of NC machining are as follows [28]:

$$
\begin{aligned}
&\text{s. t.}\frac{\pi d_0 n_{\min}}{1000} \le v_c \le \frac{\pi d_0 n_{\max}}{1000}; \quad f_{\min} \le f \le f_{\max};\ C_{F_f} a_{\text{sp}}^{x_{\text{Ff}}} f^{y_{\text{Fc}}} v_c^{n_F} K_{F_f} \le F_{\max}; \\
&\frac{F_c v_c}{1000\eta} \le P_{\max}; \quad R_a \le R_{\max}
\end{aligned}
\tag{7.50}
$$

In Eq. (7.50), n is the speed of spindle, $n_{\min}$ and $n_{\max}$ represent the minimum and maximum speeds of the NC machine tool spindle, respectively; $f_{\min}$ and $f_{\max}$ express the minimum and maximum feeds allowed by the machine; C_{Ff}, x_{Ff}, y_{Ff}, n_{Ff}, and K_{Ff} represent the coefficients related to the workpiece material and cutting conditions, which can be obtained by referring to the cutting allowance manual; F_c is the cutting force, η is the effective coefficient of the machine tool power, and $P_{\max}$ is the maximum effective cutting power of the machine tool; $R_{\max}$ is the maximum value required for the surface roughness of the part.

By using the data and coefficients from [28], the domain of the input variables is bounded in the area of ABCDE in Fig. 7.6, in which $x_1 = v_c$ and $x_2 = f$. The ratio of the area ABCDE to the corresponding virtual rectangle ABFE is 82.06%. Therefore, the uniform table $U_{21}{*}(21^7)$ is employed so as to ensure 17 sampling points sitting within the area ABCDE. Table 7.16 shows the positions of the sampling points of $U_{21}{*}(21^7)$ within domain [0.418, 7.319] × [0.1, 3.5], and actually, we have 18 sampling points sitting within the area ABCDE; see Fig. 7.6.

Shown in Table 7.17 are the discrete values of the objective functions $T_p(x_1, x_2)$ and $C(x_1, x_2)$, together with the assessed consequences for their preferable probability.

Table 7.17 indicates that the sampling point *No. 10* gets the maximum value of total preferable probability P_t; therefore, the optimal solution for this problem is

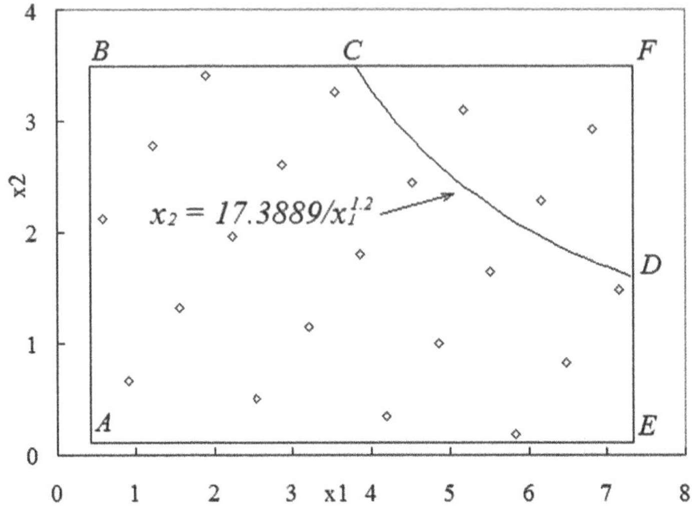

Fig. 7.6 Distribution diagram of the sampling points of $U_{21}*(21^7)$ within domain [0.418, 7.319] × [0.1, 3.5]

at $x_1 = 3.5399$ m/s and $x_2 = 3.2571$ mm/r, and the optimal values of objective functions processing time $T_p(x_1, x_2)$ and carbon emission $C(x_1, x_2)$ are 68.6078 s and 204.2189 g for the fixed cutting depth of 1 mm, respectively. The result of the total preferable probability at sampling point *No. 13* follows that of the sampling point *No. 10* tightly.

7.6 Conclusive Remarks

From the above discussion, the efficient approach with a small number of sampling points for simplifying calculation of a definite integral is useful to conduct the corresponding computation in the probability-based multi-objective optimization (PMOO), by using this efficient approach the complicated definite integral in the PMOO can be simplified as a summation of the corresponding values of the function at the small number of discrete sampling points in the integral domain. The good lattice point and uniform experimental design method are the basis of this simplification; the discrete sampling points are distributed according to the rules of good lattice points and uniform design method.

Table 7.16 Positions of the sampling points of U_{21}*(21^7) within domain [0.418, 7.319] × [0.1, 3.5]

No.	x_{10}	x_{20}	x_1 (m/s)	x_2 (mm/r)	Notice
1	1	13	0.5823	2.1238	In ABCDE
2	2	4	0.9109	0.6667	In ABCDE
3	3	17	1.2395	2.7714	In ABCDE
4	4	8	1.5682	1.3143	In ABCDE
5	5	21	1.8968	3.4190	In ABCDE
6	6	12	2.2254	1.9619	In ABCDE
7	7	3	2.5540	0.5048	In ABCDE
8	8	16	2.8826	2.6095	In ABCDE
9	9	7	3.2113	1.1524	In ABCDE
10	10	20	3.5399	3.2571	In ABCDE
11	11	11	3.8685	1.8	In ABCDE
12	12	2	4.1971	0.3429	In ABCDE
13	13	15	4.5257	2.4476	In ABCDE
14	14	6	4.8544	0.9905	In ABCDE
15	15	19	5.1830	3.0952	
16	16	10	5.5116	1.6381	In ABCDE
17	17	1	5.8402	0.1810	In ABCDE
18	18	14	6.1688	2.2857	
19	19	5	6.4975	0.8286	In ABCDE
20	20	18	6.8261	2.9333	
21	21	9	7.1547	1.4762	In ABCDE

Table 7.17 Discrete values of the objective functions $T_p(x_1, x_2)$ and $C(x_1, x_2)$, and assessed consequences

No.	Objective value		Preferable probability			Rank
	T_p/s	C/g	P_{time}	P_C	$P_t \times 10^2$	
1	240.1212	670.7583	0.0410	0.0422	0.1729	16
2	439.2724	1234.0802	0.0105	0.0112	0.0118	18
3	117.1648	332.6156	0.0598	0.0607	0.3634	12
4	163.2869	466.5245	0.0528	0.0534	0.2817	13
5	84.6379	243.6970	0.0648	0.0656	0.4253	7
6	102.4217	296.9690	0.0621	0.0627	0.3892	10
7	232.3224	677.0078	0.0422	0.0418	0.1765	15
8	79.5868	233.8848	0.0656	0.0662	0.4339	5
9	112.2090	332.0162	0.0606	0.0608	0.3682	11
10	**68.6078**	**204.2189**	**0.0673**	**0.0678**	**0.4560**	**1**
11	82.1230	246.2396	0.0652	0.0655	0.4269	6
12	213.1375	642.3436	0.0451	0.0437	0.1973	14
13	**69.4499**	**210.8595**	**0.0671**	**0.0674**	**0.4527**	**2**
14	97.4192	297.9239	0.0629	0.0626	0.3937	9
16	74.3177	230.1111	0.0664	0.0664	0.4406	4
17	272.8906	848.7030	0.0360	0.0324	0.1165	17
19	92.1370	290.8548	0.0637	0.0630	0.4013	8
21	70.4974	225.1731	0.0670	0.0666	0.4463	3

References

1. G. Leobacher, F. Pillichshammer, *Introduction to Quasi-Monte Carlo Integration and Applications* (Springer, Cham, 2014). https://doi.org/10.1007/978-3-319-03425-6
2. K.-T. Fang, M.-Q. Liu, H. Qin, Y.-D. Zhou, *Theory and Application of Uniform Experimental Designs* (Science Press/Springer, Beijing/Singapore, 2018). https://doi.org/10.1007/978-981-13-2041-5
3. K.-T. Fang, Y. Wang, *Number—Theoretic Methods in Statistics* (Chapman & Hall, London, 1994). ISBN: 0-412-46520-5
4. L.-K. Hua, Y. Wang, *Applications of Number Theory to Numerical Analysis* (Springer/Science Press, Berlin/Beijing, 1981). ISBN: 9783540103820
5. S.H. Paskov, New methodologies for valuing derivatives, in *Mathematics of Derivative Securities*, ed. by S. Pliska, M. Dempster (Isaac Newton Institute & Cambridge University Press, Cambridge, 1996), pp. 545–582. https://doi.org/10.7916/D8TB1FRJ
6. S.H. Paskov, J.F. Traub, Faster valuation of financial derivatives. J. Portf. Manag. **22**(1), 113–120 (1995). https://doi.org/10.3905/jpm.1995.409541
7. I.H. Sloan, H. Woiniakowski, When are quasi-Monte Carlo algorithms efficient for high dimensional integrals? J. Complex. **14**(1), 1–33 (1998). https://doi.org/10.1006/jcom.1997.0463
8. S. Tezuka, Financial applications of Monte Carlo and Quasi-Monte Carlo methods, in *Random and Quasi-Random Point Sets. Lecture Notes in Statistics*, vol. 138, ed. by P. Hellekalek, G.

Larcher (Springer, New York, 1998), pp. 303–332. https://doi.org/10.1007/978-1-4612-1702-27
9. S. Tezuka, Quasi-Monte Carlo—discrepancy between theory and practice, in *Monte Carlo and Quasi-Monte Carlo Methods 2000*, ed. by K.T. Fang, H. Niederreiter, F. J. Hickernell (Springer, Heidelberg, 2002), pp. 124–140. https://doi.org/10.1007/978-3-642-56046-0_8
10. R.E. Caflisch, W. Morokoff, A. Owen, Valuation of mortgage-backed securities using Brownian bridges to reduce effective dimension. J. Comput. Finance **1**(1), 27–46 (1997). https://doi.org/10.21314/JCF.1997.005
11. A. Papageorgiou, J.F. Traub, Faster evaluation of multi-dimensional integrals. Comp. Phys. **11**, 574–578 (1997). https://doi.org/10.1063/1.168616
12. B.D. Ripley, *Spatial Statistics* (Wiley, NJ, 1981). ISBN: 0-47169116-X
13. R. Dunn, A.R. Harrison, Two-dimensional systematic sampling of land use. Appl. Statist. **42**(4), 585–601 (1993)
14. J.F. Wang, A. Stein, B.B. Gao, Y. Ge, A review of spatial sampling. Spat. Stat. **2**, 1–14 (2012)
15. M.Q. Liu, D.K.J. Lin, Y. Zhou, The contribution to experimental designs by Kai-Tai Fang, in *Contemporary Experimental Design, Multivariate Analysis and Data Mining, Festschrift in Honour of Professor Kai-Tai Fang*, ed. by J. Fan, J. Pan (Springer, Cham, 2020), pp. 21–35. https://doi.org/10.1007/978-3-030-46161-4
16. W. Navidi, *Statistics for Engineers and Scientists*, 5th edn. (McGraw-Hill Education, New York, 2020). ISBN: 9781260431025
17. S. Ju, X. Yang, G. Liu, Application of elliptic integral calculus in computing magnetic induction intensity. J. Changchun Inst. Technol. (Nat. Science. Ed.). **6**, 70–72 (2005). https://kns.cnki.net/kcms/detail/detail.aspx?dbcode=CJFD&dbname=CJFD2005&filename=CGCZ200504024&uniplatform=NZKPT&v=ryxpXwQqLAxChu6YS1miYma1Skh6baZ6482KF04E5VURFrx0iFF1b9u19uS2b7G2 (in Chinese)
18. P.F. Byrd, M.D. Friedman, *Hand Book of Elliptic Integrals for Engineers and Scientists*, 2nd edn. (Springer, Berlin, Heidelberg, 1971). https://doi.org/10.1007/978-3-642-65138-0
19. S. Song, S. Chen, Two effective quadrature schemes for calculating double integration. J. Zhengzhou Univ. **36**, 16–19 (2004) https://kns.cnki.net/kcms/detail/detail.aspx?dbcode=CJFD&dbname=CJFD2004&filename=ZZDZ200401003&uniplatform=NZKPT&v=kCi42ClkKjv8yV3dacrXPRtdvn0JHnYAx3nBIo6p2qVlawH9fBMWaaiKoulEFMmf
20. Z. Chen, X. Zheng, C. Luo, Y. Zhang, X. Chen, Integration of multivariate functions by orthogonal arrays. J. Shanghai Inst. Technol. (Nat. Sci. Ed.) **2**, 119–123 (2010). http://caod.oriprobe.com/articles/25690087/integration_of_Multivariate_Functions_by_Orthogonal_Arrays.htm
21. H. Zheng, J. Hu, X. Li, X. Cao, Monte Carlo methods of high-dimensional numerical integration. J. Nanchang Hangkong Univ. **23**, 37–41 (2009)
22. J. Han, W. Ren, Monte Carlo integration and quasi-Monte Carlo integration. J. Shanxi Normal Univ. (Nat. Sci. Ed.) **21**, 13–17 (2007)
23. J. Yu, M. Zheng, Y. Wang, H. Teng, An efficient approach for calculating a definite integral with about a dozen of sampling points. Vojnotehnički glasnik/Mil. Tech. Courier **70**(2), 340–356 (2022). https://doi.org/10.5937/vojtehg70-36029
24. B. Huang, D. Chen, Effective Pareto optimal set of multi-objective optimization problems. Comput. Digit. Eng. **37**(2), 28–34 (2009)
25. X. Qu, N. Lu, X. Meng, Multi-objective fuzzy optimization of tower crane boom tie rods. Mech. Transm. **28**(3), 38–40 (2004). 1004-253912004)03-0038-03
26. Q. Gao, D. Pan, Random uniform grid and its application in rubber description optimization problems. Syst. Eng. Theor. Pract. **19**(11), 87–91 (1999)
27. Q. Xiong, Study of treatment of multi-objective problem on basis of linear programming model. Sci. Technol. Innov. **25**(28), 42–43 (2020)
28. C. Li, L. Cui, F. Liu, L. Li, Multi-objective NC machining parameters optimization model for high efficiency and low carbon. J. Mech. Eng. **49**(9), 87–96 (2013). https://doi.org/10.3901/JME.2013.09.087

Chapter 8
Fuzzy-Based Probabilistic Multi-objective Optimization for Material Selection

Abstract In this chapter, the probability-based method for multi-objective optimization (PMOO) is adopted to develop a rational fuzzy-related multi-objective optimization for material selection, the fundamental idea and algorithm of fuzzy as well as probability theory are taken as the footstone to conduct the formulation. The utility of the material performance index is determined by the intersection of the membership function of fuzzy numbers of candidate material performance and the membership function of fuzzy numbers of desired material performance. Thereafter, the utility of each material performance index is employed to conduct the evaluation of partial preferable probability and formulate the multi-objective optimization by means of probability theory. Furthermore, a typical example is given to illustrate the rational process of fuzzy multi-objective optimization for material selection.

8.1 Introduction

Multi-objective optimization (MOO) is quite useful in solving problems involving the evaluation of multiple attributes and alternatives [1–4]. It is quite often that the objectives (attributes) are conflict with each other, which makes it difficult to provide a rational decision without systemic consideration [5–8].

On the other hand, in some cases a linguistic or approximated expression for responses is available, which makes the evaluation with characteristic of "Fuzzy" in some sense [1–4].

Appropriate material selection for structures or machine components is one of the most complicated and time-consuming problems for engineers or manufacturing companies, since it involves many conflicting objectives and feasible alternatives, which is even with characteristic of "Fuzzy." The determination and evaluation of both beneficial and unbeneficial types of attributes of one alternative are all involved intricately.

M. Zheng et al., *Probability-Based Multi-objective Optimization for Material Selection*, https://doi.org/10.1007/978-981-99-3939-8_8

Many potential attributes (objectives) for material selection, such as hardness, machinability, cost, and corrosive resistance, must be considered together in the material selection process. Therefore, material selection for structures or machine components can be viewed as a multi-objective optimization (MOO) problem undoubtedly, especially in case of the presences of many quantitative and qualitative criteria [5–8].

Under condition of presences of quantitative and qualitative criteria, a designer has to face the problem of choosing an appropriate quantified method to deal with non-quantifiable criteria. The evaluation concerning qualitative criteria is usually subjective and thus imprecise [1–8]. Therefore, it is very usual to develop a comparative criterion that can be employed to describe the preference of one alternative over others [9, 10]. Some evaluation criteria for alternative suitability of material selection are subjective ones or in linguistic terms, such as the weight importance of the attribute, and the corrosive resistance, which makes it an actual fuzzy problem.

Many investigators attempted to develop fuzzy MOO methods for material selection. Most of them deal with the problems by simply combining the fuzzy concept with traditional MOO approaches [1–10], such as the "Technique for Order Preference by Similarity to Ideal Solution" (TOPSIS), "analytic hierarchy process" (AHP), "VIšekriterijumska Optimizacija I KOmpromisno Resenje" (VIKOR), and "multi-objective optimization on basis of ratio analysis" (MOORA). However, this kind of combination of fuzzy concept with traditional MOO approaches could not be viewed as reasonable hybrids, since there exist inherent shortcomings in the traditional MOO approaches with their "additive algorithm" and scaled factor for various objectives [11–14].

Currently, a probability-based method for multi-objective optimization (PMOO) was developed, which aims to solve the intrinsic problems of the traditional MOO with "additive" algorithms [11–14]. The novel idea of preferable probability was proposed to reflect the preferable degree of a performance utility indicator in the optimization. In the novel methodology, all performance utility indicators of alternatives were preliminarily categorized as two types, i.e., both beneficial or unbeneficial types according to their specific functions and preference in the optimization; each performance utility indicator of the alternative makes its contribution as a partial preferable probability to the entire optimization quantitatively; furthermore, according to probability theory and set theory, the product of all partial preferable probabilities produces the overall preferable probability of the alternative, which is the uniquely deterministic index in the optimization process and thus converts the multi-objective optimization problem into a mono-objective optimization one. The simultaneous optimization of all performance utility indicators is the intrinsic issue of multi-objective optimization, this matter is rationally reflected by the product of all partial preferable probabilities equaling to the total preferable probability of an alternative in the spirit of probability theory [11–14].

In this chapter, the probability-based method for multi-objective optimization (PMOO) is adopted to formulate a rational hybrid optimization of fuzzy concept with MOO methods for material selection, which takes the fundamental idea and algorithm of fuzzy as well as probability theory as the starting point to perform the formulation of fuzzy probability-based multi-objective optimization (FPMOO). Furthermore, one example is given to illustrate the rational process of FPMOO for material selection.

8.2 Formulation of Fuzzy Probability-Based Multi-objective Optimization (FPMOO)

8.2.1 *Membership Value of Material Performance in Fuzzy Language*

As has been mentioned that there exist many quantitative and qualitative criteria material selection problems usually, since the deterministic evaluation of quantitative criteria in multi-objective optimization (MOO) for material selection is conducted in [11–13], here in this chapter only the evaluation of quantitative and qualitative criteria with fuzzy characteristic in multi-objective optimization (MOO) problem is concerned analogically.

(1) **Membership of quantitative performance**

This kind of material performance can be usually expressed by numerical data. However, since the operations of material processing are with stochastic nature, such kind of material performances are not well fixed but ranging in some areas (lower limit to upper limit). For example, the hardness of stainless steel 410 is [155, 350] HB approximately [3]. Such kind of quantitative performances can be categorized as fuzzy number of "*1st Type.*" Sometimes, the lower limit and upper limit of the range interval are uncertain either. In this case, the quantitative properties can be seen as fuzzy number of "*2nd Type.*"

Since the value of a quantitative performance of material can be ranging in an interval (between lower limit and upper limit) approximately, a trapezoidal function can be employed to reflect the membership value of the quantitative performances of material [3]. For example, hardness of stainless steel 410 is ranging in "155–350" approximately, which can be subjectively represented by (139, 155, 350, 385) suffering 10% fuzziness in its property value in the database in lower limit and upper limit [3], the membership function of this kind of quantitative performance is shown by Form 2 of Fig. 8.1. On the other hand, a property with a value "approximately equal to 200" can be represented by (180, 200, 200, 220), which is shown by Form 4 in Fig. 8.1. Analogically, Form 1 and Form 3 gain their specific meaning of smaller or equal to same data and bigger or equal to same data [3].

(2) **Membership of qualitative performance**

A qualitative property is a response with linguistic characteristic, which is expressed in words or sentences, such as, for "corrosion resistance" the usual linguistic expression is "recommended" ($\tilde{R}_e$), or "acceptable" ($\tilde{A}_c$), or "not recommended" ($\tilde{N}\tilde{R}$);

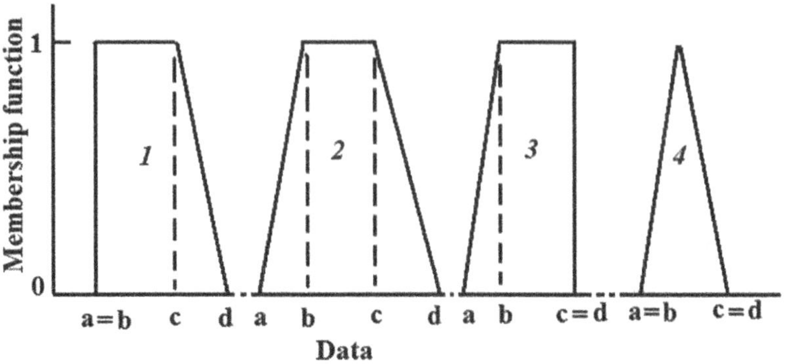

Fig. 8.1 Forms of trapezoidal membership function

for importance weight of the relevant material properties the common linguistic demonstration is "high" ($\tilde{w}_1$), or "very high" ($\tilde{w}_2$), or "medium" ($\tilde{w}_3$), etc.

As to material property and importance weight, they actually belong to different categories.

The importance weight can be evaluated by subjective score, for example, the importance weight of the relevant material properties, with "high" ($\tilde{w}_1$) can be scored by "8," the "very high" ($\tilde{w}_2$) can be scored by "10," and the "medium" ($\tilde{w}_3$) can be scored by "6."

On the other hand, the qualitative performance of material property in fact has certain meaning, which can be expressed by trapezoidal fuzzy numbers as well, for example, according to guidelines for classification given in Table 8.1 [3], the membership functions for the "corrosion resistance" were subjectively defined as follows: $\tilde{R}_e$: (18, 18, 18, 22); $\tilde{A}_c$: (18, 20, 50, 55); $\tilde{N}\tilde{R}$: (45, 55, 55, 55).

(3) **Desired data and available data of material performances**

The desired data of material performances for design can also be transferred into trapezoidal fuzzy numbers as a requirement for material selection. For example, the desired value of Brinell hardness equals to 300 HB approximately ($\tilde{D}_1$), and it can be represented by trapezoidal fuzzy numbers as (270, 300, 300, 330), which is reflected by Form 4 in Fig. 8.1.

Usually, the data of material performances from practical production or handbook is used as available data to withstand the screening of the material selection.

(4) **Utility of material performance**

The "intersection" of the "desired data" and "available data" of material performances can be used to determine the utility of material performance in the material selection.

(A) Under condition of the range of the desired trapezoidal fuzzy numbers fully covering the available trapezoidal fuzzy numbers of material performances, the consequence of the utility of the corresponding material property is "1."

Table 8.1 Data of some metallic material performance properties [3]

Metal	Hardness (HB)	Machinability rating* (%)	Cost ($/lb)	Corrosion resistance⊕
Stainless steel 17-4PH	270–420	25	4–5	Recommended
Stainless steel 410	155–350	40	3	Recommended
Stainless steel 440A	215–390	30	2.5–3.0	Recommended
Stainless steel 304	150–330	45	2	Not recommended
Ni-resist cast iron	130–250	35	0.8–1.3	Recommended
High-chromium cast iron	250–700	25	2–2.5	Recommended
Ni-hard cast iron	525–600	30	1.8–2.2	Recommended
Nickel 200	75–230	55	4	Acceptable
Monel 400	110–240	35	8	Recommended
Inconel 600	170–290	45	8.5–9.0	Recommended

Notes * The machinability rating of Cold-drawn AISI 1112 steel is taken as a value of 100%; ⊕ the guidelines state: "recommended" for corrosion rate < 0.02 mm/a, "acceptable" for 0.02 < corrosion rate < 0.05 mm/a, and "not recommended" for corrosion rate > 0.05 mm/a

(B) In case of the desired trapezoidal fuzzy numbers covering nothing related to the available trapezoidal fuzzy numbers of material performances, the consequence of the utility of the corresponding material property is "0."

(C) Otherwise, if the desired trapezoidal fuzzy numbers partially covering part of the available trapezoidal fuzzy numbers of material performances, the consequence of the utility of the corresponding material property is the ratio of the area of the covered part to the total area of the available trapezoidal fuzzy numbers of material performances.

For example, the desired value of Brinell hardness is equaled to 300 HB approximately ($\tilde{D}_d$), its trapezoidal fuzzy numbers can be represented as $\tilde{D}_d$: (270, 300, 300, 330), the available trapezoidal fuzzy numbers of Nickel 200 are given by $\tilde{D}_a$: (67, 75, 230, 253), it is obvious that there is no "intersection" between $\tilde{D}_d$ and $\tilde{D}_a$ for Nickel 200; therefore, the utility of Nickel 200 in hardness is "0"; while the available trapezoidal fuzzy numbers of stainless steel 410 are given by $\tilde{D}_a$: (139, 155, 350, 385), there is a "intersection" between $\tilde{D}_d$ and $\tilde{D}_a$ for stainless steel 410 in hardness, and the result of utility of stainless steel 410 in hardness is 0.1354. Besides, the desired value of cost is smaller or equal to $ 3.5/lb approximately ($\tilde{D}_d$), its trapezoidal fuzzy numbers can be represented as $\tilde{D}_d$: (0, 0, 3.5, 3.85), the available trapezoidal fuzzy numbers of stainless steel 410 are given by $\tilde{D}_a$: (2.7, 3, 3, 3.3), the trapezoidal fuzzy numbers of $\tilde{D}_a$ are fully covered by the trapezoidal fuzzy numbers of $\tilde{D}_d$ for stainless steel 410, and therefore the utility of stainless steel 410 in cost is "1."

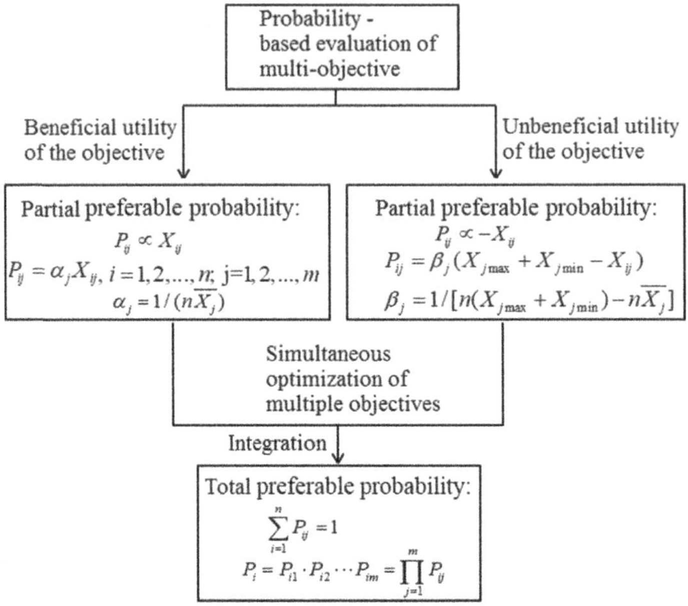

Fig. 8.2 Procedure of PMOO method

8.2.2 Fuzzy Probability-Based Multi-objective Optimization (FPMOO)

As the utility of available material performance is evaluated by the procedures of last sections in the respect of fuzzy language and set theory, which is withstanding the screening of desired indexes, the available data of material performance can be thus used to the conduct fuzzy probability-based multi-objective optimization (FPMOO) rationally [11–14].

Figure 8.2 shows the general procedure of the "probability-based multi-objective optimization" from utility of material performance indexes.

8.3 Illustrative Example

Take the engineering application of material selection for a nozzle of a jet fuel system as an example [3]. The specific desired values and importance weight of the material are cited in Table 8.2 [3], which is used to screen the available candidate materials.

The general meanings of Table 8.2 are as followings, the Brinell hardness equal to 300 HB approximately ($\tilde{D}_1$), the machinability rating is greater or equal to 30 approximately ($\tilde{D}_2$), the cost is smaller or equal to \$3.5/lb approximately ($\tilde{D}_3$), and corrosion resistance is as "recommended" ($\tilde{D}_4$). The evaluation of the importance

Table 8.2 Desired values material properties

Property	Hardness (HB), $\tilde{D}_{d1}$	Machinability rating* (%), $\tilde{D}_{d2}$	Cost ($/lb), $\tilde{D}_{d3}$	Corrosion resistance, $\tilde{D}_{d4}$
$\tilde{D}_d$	(270, 300, 300, 330)	(27, 30, 100, 100)	(0, 0, 3.5, 3.85)	(18, 18, 18, 22)
Importance weight	8	10	6	6

weight of the desired material property is scored by "8" for "high" ($\tilde{w}_1$) for hardness, "10" for "very high" ($\tilde{w}_2$) for machinability rating, and "6" for "medium" for cost ($\tilde{w}_3$) and for corrosion resistance ($\tilde{w}_4$).

The candidate materials are shown in Table 8.1, which is converted into the trapezoidal fuzzy numbers and thus presented in Table 8.3. The fuzzy numbers were determined based on the subjective hypothesis that there is 10% fuzziness in each property value of the database.

Table 8.4 shows the utility of the candidate materials by means of screening with the desired requirement for the material selection. Table 8.5 shows the consequence of evaluation for preferable probability and rank. In the evaluation of partial preferable

Table 8.3 Trapezoidal fuzzy numbers of candidate material properties corresponding to Table 8.1

Metal	Hardness, $\tilde{D}_{a_{i1}}$	Machinability rating, $\tilde{D}_{a_{i2}}$	Cost, $\tilde{D}_{a_{i3}}$	Corrosion resistance, $\tilde{D}_{a_{i4}}$
Stainless steel 17-4PH	(243, 270, 420, 462)	(22, 25, 25, 28)	(3.6, 4, 5, 5.5)	(18, 18, 18, 22)
Stainless steel 410	(139, 155, 350, 385)	(36, 40, 40, 44)	(2.7, 3, 3, 3.3)	(18, 18, 18, 22)
Stainless steel 440A	(188, 215, 390, 429)	(27, 30, 30, 33)	(2.2, 2.5, 3, 3.3)	(18, 18, 18, 22)
Stainless steel 304	(140, 150, 330, 363)	(40, 45, 45, 50)	(1.8, 2, 2, 2.2)	(45, 55, 55, 55)
Ni-resist cast iron	(117, 130, 250, 275)	(31, 35, 35, 39)	(0.7, 0.8, 1.3, 1.4)	(18, 18, 18, 22)
High-chromium cast iron	(225, 250, 700, 770)	(22, 25, 25, 28)	(1.8, 2, 2.5, 2.8)	(18, 18, 18, 22)
Ni-hard cast iron	(472, 525, 600, 660)	(27, 30, 30, 33)	(1.6, 1.8, 2.2, 2.4)	(18, 18, 18, 22)
Nickel 200	(67, 75, 230, 253)	(49, 55, 55, 61)	(3.6, 4, 4, 4.4)	(18, 20, 50, 55)
Monel 400	(100, 110, 240, 264)	(31, 35, 35, 39)	(7.2, 8, 8, 8.8)	(18, 18, 18, 22)
Inconel 600	(153, 170, 290, 319)	(40, 45, 45, 50)	(7.6, 8.5, 9, 9.9)	(18, 18, 18, 22)

probability of the utility of material performance index, all above 4 utilities have the characteristic of "the higher the better," so they all belong to beneficial type of indexes, as to the evaluation of total preferable probability of the candidate material, total preferable probability is expressed by [11–14]

$$P_t = P_{hd}^{w1} \times P_{mr}^{w2} \times P_c^{w3} \times P_{cr}^{w4}, \quad (8.1)$$

the symbols P_{hd}, P_{mr}, P_c, and P_{cr} in Eq. (8.1) represent the partial preferable probability of hardness, machinability, cost, and corrosion resistance, respectively, w_1, w_2, w_3, and w_4 indicate the corresponding importance weight individually.

It can be seen from Table 8.5 that stainless steel 440A is ranked first, which is closely followed by stainless steel 410.

Table 8.4 Utility of the candidate materials by means of screening with the desired requirement for the material selection

Metal	Hardness (hd), $\tilde{U}_{a_{i1}}$	Machinability rating (mr), $\tilde{U}_{a_{i2}}$	Cost (c), $\tilde{U}_{a_{i3}}$	Corrosion resistance (cr), $\tilde{U}_{a_{i4}}$
Weight importance	8	10	6	6
Normalized weight importance	0.2667	0.3333	0.2	0.2
Stainless steel 410	0.1354	1	1	1
Stainless steel 440A	0.1442	1	1	1
Stainless steel 304	Unavailable			
Ni-resist cast iron	0.0016	1	1	1
High-chromium cast iron	0.0603	0.0278	1	1
Ni-hard cast iron	Unavailable			
Nickel 200	Unavailable			
Monel 400	Unavailable			
Inconel 600	Unavailable			

Table 8.5 Consequence of evaluation for preferable probability and rank

Metal	Partial preferable probability				Total preferable probability	Rank
	P_{hd}	P_{mr}	P_c	P_{cr}	P_t	
Stainless steel 410	0.3965	0.3303	0.25	0.25	0.3102	2
Stainless steel 440A	0.4223	0.3303	0.25	0.25	0.3155	1
Ni-resist cast iron	0.0047	0.3303	0.25	0.25	0.0950	3
High-chromium cast iron	0.1766	0.0092	0.25	0.25	0.0757	4

8.4 Concluding Remarks

From the above discussion, it can be seen that the fuzzy multi-objective optimization on basis of probability theory for material selection is well-proposed. It is reasonably to determine the utility of the material performance index by using the intersection of the membership function of fuzzy numbers of candidate material performance and the membership function of fuzzy numbers of desired material performance. The utility of each material performance index can be naturally used to formulate the fuzzy multi-objective optimization in respect of probability theory.

References

1. M. Enea, T. Piazza, Project selection by constrained fuzzy AHP. Fuzzy Optim. Decis. Making **3**, 39–62 (2004)
2. S. Önüt, S. Soner Kara, T. Efendigil, A hybrid fuzzy MCDM approach to machine tool selection. J. Intell. Manuf. **19**, 443–453 (2008). https://doi.org/10.1007/s10845-008-0095-3
3. T.W. Liao, A fuzzy multicriteria decision-making method for material selection. J. Manuf. Syst. **15**(1), 1–12 (1996)
4. P. Vasant, N.N. Barsoum, Fuzzy optimization of units products in mix-product selection problem using fuzzy linear programming approach. Soft Comput. **10**, 144–151 (2006). https://doi.org/10.1007/s00500-004-0437-9
5. M. Babanli, T. Gojayev, Application of fuzzy AHP method to material selection problem, in *WCIS 2020, AISC 1323*, ed. by R.A. Aliev et al. (Springer, Cham, 2021), pp. 254–261. https://doi.org/10.1007/978-3-030-68004-6_33
6. G. Vats, R. Vaish, Piezoelectric material selection for transducers under fuzzy environment. J. Adv. Ceram. **2**(2), 141–148 (2013). https://doi.org/10.1007/s40145-013-0053-1
7. I.V. Germashev, M.A. Kharitonov, E.V. Derbisher, V.E. Derbisher, Selection of components of a composite material under fuzzy information conditions, in *Cyber-Physical Systems: Advances in Design & Modelling, Studies in Systems, Decision and Control 259*, ed. by A.G. Kravets et al. (Springer, Cham, 2020). https://doi.org/10.1007/978-3-030-32579-4_17
8. Y. Deng, Plant location selection based on fuzzy TOPSIS. Int. J. Adv. Manuf. Technol. **28**, 839–844 (2006). https://doi.org/10.1007/s00170-004-2436-5
9. A. Dikshit-Ratnaparkhi, D. Bormane, R. Ghongade, A novel entropy-based weighted attribute selection in enhanced multicriteria decision-making using fuzzy TOPSIS model for hesitant fuzzy rough environment. Complex. Intell. Syst. **7**, 1785–1796 (2021). https://doi.org/10.1007/s40747-020-00187-8
10. M. Tavana, A. Shaabani, F.J. Santos-Arteaga, N. Valaei, An integrated fuzzy sustainable supplier evaluation and selection framework for green supply chains in reverse logistics. Environ. Sci. Pollut. Res. **28**, 53953–53982 (2021). https://doi.org/10.1007/s11356-021-14302-w
11. M. Zheng, Y. Wang, H. Teng, A novel method based on probability theory for simultaneous optimization of multi-object orthogonal test design in material engineering. Kovove Mater. **60**(1), 45–53 (2022). https://doi.org/10.31577/km.2022.1.45
12. M. Zheng, Y. Wang, H. Teng, A novel approach based on probability theory for material selection. Materialwiss. Werkstofftech. **53**(6), 666–674 (2022). https://doi.org/10.1002/mawe.202100226

13. M. Zheng, H. Teng, J. Yu, Y. Cui, Y. Wang, *Probability-Based Multi-objective Optimization for Material Selection* (Springer, Singapore, 2022). https://link.springer.com/book/9789811933509
14. M. Zheng, J. Yu, A probability-based fuzzy multi-objective optimization for material selection. Technicki Glas. **18**(2), 153–158 (2024). https://doi.org/10.31803/tg-20230515054622

Chapter 9
Cluster Analysis of Separation of "Independent Objective" for Probability-Based Multi-objective Optimization

Abstract In the evaluation of the probability-based multi-objective optimization, the need of "independent response" of the multi-objective is necessary, and the "independent response" is analogical as an "independent event" from the respect of probability theory. This is a novel approach with brand new concept of "preferable probability." The novel methodology can be used in many fields, including material selection, programming problem, energy planning, operation research, etc. In this chapter, the separation of an "independent objective" from "multiple objectives" is discussed by utilizing cluster analysis. The results show that using linear correlation coefficient to distinguish the "similarity" of performances is a rational method, which thus can be used to supply appropriate objective classification for the probability-based multi-objective optimization.

9.1 Introduction

In the probability-based multi-objective optimization [1], it needs "independent response" to be analogical as an "independent event" from the respect of probability theory, which thus develops a novel approach with brand new concept of "preferable probability" and the assessments. The new methodology can be used in many fields, including material selection, programming problem, energy planning, operation research, etc.

The main idea of PMOO is that the intrinsic essence of "multi-objective optimization" is "simultaneous optimization of multiple objectives" from the point of view of system theory, the probability-based method was further proposed in respect of probability theory and set theory, which considers each objective as an "independent event" analogically.

From the perspectives of probability theory and set theory, the intersection of independent events and the joint probability of independent events can be used to characterize the simultaneous appearance of multiple independent events. In this way, when it analogically equates each objective with an independent event, the problem of simultaneous optimization of multiple objectives becomes "rule-based". Of course, the equating of "each objective" to an "independent event" relies on

M. Zheng et al., *Probability-Based Multi-objective Optimization for Material Selection*, https://doi.org/10.1007/978-981-99-3939-8_9

the separation "independent events" from "multiple objectives" completely, such that the probability-based multi-objective optimization methodology can be utilized rationally.

This chapter discusses the problem of separation of "independent event" from "multiple objectives" by using cluster analysis.

Cluster analysis refers to the analysis process of grouping the collection of physical or abstract objects into multiple classes composed of similar objects. It is an important human behavior. The goal of cluster analysis is to classify things on the basis of similarity through collecting data. Clustering comes from many fields, including biology, mathematics, computer science, statistics, and economics. In different application fields, many clustering techniques have been developed, which are used to describe data, measure the similarity between different data sources, and classify data sources into different clusters.

Generally speaking, there are different degrees of similarity (affinity) between the samples or indicators studied. Therefore, according to a number of observed indicators of a batch of samples, some statistics that can measure the similarity between samples or indicators are found out, and these statistics are used as the basis for classification. The basic idea of classification is to aggregate some samples (or indicators) with greater similarity into one category and others with lesser similarity into another category. Its similarity is defined by distance or similarity coefficient. The criterion of class merging is to minimize intra-class differences and maximize inter-class differences.

The main characteristics of cluster analysis are: It is unnecessary to know the classification structure of classified objects in advance, but only a batch of geographical data; then the classified statistics are selected and calculated according to certain procedures; finally, a complete classification system diagram can be obtained naturally and objectively.

In cluster analysis, there are two types of divisions usually, i.e., Q-type cluster analysis and R-type cluster analysis according to different classified objects. R-cluster analysis is to classify attributes (responses or indicators), and Q-cluster analysis is to classify samples [2].

The main functions of R-cluster analysis are as follows: (1) it not only shows the closeness of the relationship between individual attributes (responses or indicators), but also the closeness of the combination of variables; (2) the main attributes can be selected for regression analysis according to the classification results of attributes and their relationships. While the advantages of Q-cluster analysis include: (1) the information of multiple attributes can be comprehensively used to classify samples; (2) the classification result is intuitive, and the cluster pedigree diagram clearly shows its numerical classification result; 3. the results of cluster analysis are more detailed, comprehensive, and reasonable than traditional classification methods.

Cluster analysis was initiated in biology as a branch [3]. By classifying things, understanding of complicated problems can be recognized gradually. Preliminarily, the classification methods are mostly implemented by experience or professional knowledge, so they can be classified as qualitative methods. Later, mathematical methods were adopted to get a certain quantitative classification with characteristic values gradually. It should be a development trend to use mathematical methods for quantitative and scientific classification. Cluster analysis is also an important branch of multivariate analysis, which develops rapidly. In nowadays, cluster analysis has been widely used in biology, archaeology, geology, agricultural and industrial productions, medicine, weather forecast, medicine, and other fields. Actually, in the process of cluster analysis, the class is not given in advance, but it needs to be determined according to the characteristics of the observed data, and there is unnecessary to make any assumptions about the number and structure of classes. In the clustering results, objects attributing to the same class tend to be similar to each other in a sense, while objects attributing to different classes tend to be dissimilar. The purpose of clustering analysis is to classify objects into several classes according to certain rules [3].

According to the degree of similarity, the samples (or performances) are classified one by one, and the closely related classes are clustered into a small taxon and then gradually expanded, so that the alienated ones are clustered into a large taxon, until all the samples (or performances) are clustered, forming a cluster diagram that represents the affinity. Classify the samples (or performances) according to some requirements in turn [2].

The general respect of classification is that the closer the similarity of performances is, the closer their similarity coefficient is to 1 or -l, while the similarity coefficient of unrelated performances is closer to 0.

Those that are similar are classified into one category, and those that are not are classified into different categories. The distance is the "space" characteristic between "points." Each sample is regarded as a point in P-dimensional space, and the distance between points is measured by some kind of measurement. The points that are closer to each other belong to one category, while the points that are farther away belong to different categories.

This chapter mainly discusses R-type cluster analysis. Special attention is paid to appropriate objective classification of the multiple objectives in the probability-based multi-objective optimization [4].

9.2 Characterization of Similarity Between Performances or Samples

(A) Similarity coefficient

Usually, the similarity coefficient is used to characterize the similarity between performances or samples. But in the following section, it can be seen that the similarity coefficient is not a good coefficient to characterize the similarity between samples or performances through careful analysis of the similarity coefficient.

The definition of "similarity coefficient" is,

$$S_{jk} = \frac{\sum_{i=1}^{m} x_{ij} x_{ik}}{\left[\sum_{i=1}^{m} x_{ij}^2 \cdot \sum_{i=1}^{m} x_{ik}^2\right]^{0.5}}. \tag{9.1}$$

In Eq. (9.1), S_{jk} is a "similarity coefficient" that was used to represent the "similarity degree" between two attributes x_{ij} and x_{ik} in time or space previously.

Furthermore, the attribute x_{ij} can also be "normalized," that is,

$$y_{ij} = (x_{ij} - A_j)/(B_j - A_j). \tag{9.2}$$

In Eq. (9.2), $A_j = \min\{x_{ij}, i = 1, 2, \ldots, m\}$, $B_j = \max\{x_{ij}, i = 1, 2, \ldots, m\}$. Thereafter, it derives a "normalized expression" for Eq. (9.1),

$$S'_{jk} = \frac{\sum_{i=1}^{m} y_{ij} y_{ik}}{\left[\sum_{i=1}^{m} y_{ij}^2 \cdot \sum_{i=1}^{m} y_{ik}^2\right]^{0.5}}. \tag{9.3}$$

In fact, the results of Eq. (9.1) and Eq. (9.3) are not equivalent in general. In addition, the results of Eq. (9.1) and Eq. (9.3) can't even reflect the similarity between two diagraphs, which can be seen in the following example. Therefore, the "similarity coefficient" between two attributes obtained from above definition may not be a "good coefficient" to characterize the real similarity between performances or samples.

Example 1 Table 9.1 gives the values of three attributes of seven different samples. Curves 1, 2, and 3 in Fig. 9.1 show the variations characteristics of these three attributes of different samples, respectively. Now, the problem is to find the "similarity coefficient" between attribute (1) and attributes (2) and (3).

According to Eq. (9.1), the "similarity coefficient" between attribute 1 and attribute 2 of these seven samples can be obtained,

$$S_{12} = 0.9829. \tag{9.4}$$

While, according to Eq. (9.3), the "similarity coefficient" between attribute 1 and attribute 2 of these seven samples can be obtained,

Table 9.1 Values of three attributes for seven samples

Sample	Attribute 1	Attribute 2	Attribute 3
1	1	4	10
2	2	5	9
3	3	6	8
4	4	7	7
5	5	8	6
6	6	9	5
7	7	10	4

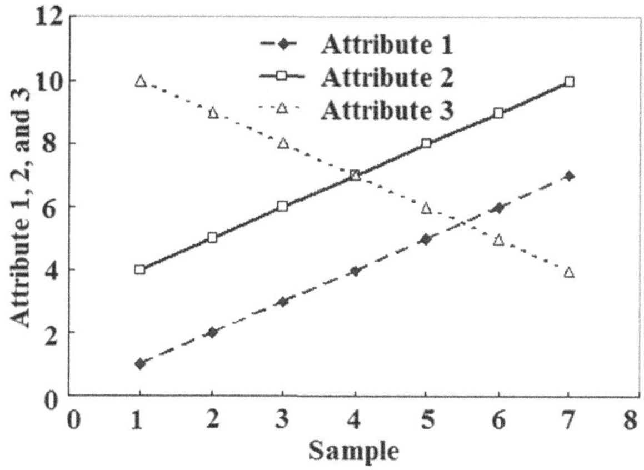

Fig. 9.1 Three attributes of seven samples

$$S'_{12} = 1. \tag{9.5}$$

The value given by Eq. (9.4) does not equal to that of Eq. (9.5) for the same attributes! This shows that the "similarity coefficient" defined in such a way cannot be seen as "good coefficient" to characterize the similarity between attributes.

Furthermore, let's study the "similarity coefficient" between attribute 1 and attribute 3 to see what happens.

According to Eq. (9.1), the "similarity coefficient" between attribute 1 and attribute 3 of these seven samples is,

$$S_{13} = 0.7372. \tag{9.6}$$

However, according to Eq. (9.3), the similarity coefficient between attribute 1 and attribute 3 of these seven samples is,

$$S'_{13} = 0.3846. \tag{9.7}$$

These results completely show that the "similarity coefficient" defined in above way is not a proper one to characterize the similarity between attributes.

There are also many articles that showed the puzzles of "similarity coefficient" [5–8].

(B) **Linear correlation coefficient**

The definition of "linear correlation coefficient" is,

$$r_{jk} = \frac{\sum_{i=1}^{m} (x_{ij} - \overline{x_j}) \cdot (x_{ik} - \overline{x_k})}{\left[\sum_{i=1}^{m} (x_{ij} - \overline{x_j})^2 \cdot \sum_{i=1}^{m} (x_{ik} - \overline{x_k})^2\right]^{0.5}}. \tag{9.8}$$

In Eq. (9.8), r_{jk} is the "linear correlation coefficient" between two attributes x_{ij} and x_{ik}, which is used to represent the degree of linear correlation degree of these two attributes; $\overline{x_j}$ is the average value of the j-th attribute and $\overline{x_k}$ the average value of the k-th attribute.

Now, the "linear correlation coefficient" is analyzed with an example as well.

Example 2 Find the linear correlation coefficient between the attributes (1) and (2) of different samples given in Fig. 9.1 and Table 9.1.

Here, we use Eq. (9.8) to analyze the data of attribute (1) and attribute (2) in Table 9.1 and Fig. 9.1. It results in the linear correlation coefficient $r_{12} = 1$, while the analysis of the data of attribute (1) and attribute (3) in Table 9.1 and Fig. 9.1 gets the linear correlation coefficient $r_{13} = -1$.

In fact, there is only a relative translation between the data of attribute (1) and attribute (2) in Table 9.1 and Fig. 9.1, that is, "*the data of attribute* 1" = "'*the data of attribute* $2' + 3''$ only. Of course, they are completely linearly correlated and fully similar.

However, according to the definition for "similarity coefficient" with Eq. (9.1), its similarity has been "discounted." So it can be said that "similarity coefficient" may not be a "good coefficient" to characterize the similarity between performances or samples. Actually, the linear correlation coefficient is just the right coefficient to reflect the linear proportional relationship, and it is more reasonable to reflect the similarity between attributes or samples; in addition, the linear correlation coefficient also has the invariance of "normalization" similar to the Eq. (9.2).

Besides, Yan et al. analyzed the precipitation forecast business and field test of Zhengzhou Meteorological Observatory during the main flood season of Huanghuai in June–August, 2000, and showed that the forecast method with linear correlation coefficient is very useful in precipitation similarity forecast, especially in 24-h forecast [9].

(C) **Distance**

The distance between samples is defined. The smaller the distance, the closer they are. The commonly used distances include Minkowski distance, Euclidean distance, and Chebyshev distance. In consideration of the dimensional differences between different attributes, the normalization like Eq. (9.2) is usually used as well.

9.3 Application of Clustering Analysis in Separation of "Independent Objective" for Multi-objective Optimization

As mentioned earlier, in the probability-based multi-objective optimization, equating "each objective" to an "independent event" depends on the separation of "independent event" from "multiple objectives" strongly. This section discusses the separation of "independent goal" from multiple objectives by using cluster analysis. The illustration is still conducted with appropriate example.

Example 3 Selection of cutting tool material

As proper tool material, which is expected to have higher flexural strength, thermal conductivity, and heat-resistant temperature, but lower coefficient of thermal expansion.

Cao once used cluster analysis to analyze the physical and mechanical properties of 8 cutting tool materials [10]. Now, it is re-analyzed to fully understand the co-relationship between the physical and mechanical properties of cutting tool materials. Table 9.2 shows the physical and mechanical properties of the eight tool materials [10].

The analysis of the data in Table 9.2 shows that there is a strong linear correlation among the hardness, flexural strength, and impact toughness of cutting tool materials. As shown in Fig. 9.2, Fig. 9.3 and Fig. 9.4, their linear correlation coefficients are greater than 94%, the flexural strength and impact toughness are positive correlated.

As to attributes of flexural strength and impact toughness of such group of cutting tool materials, when doing multi-objective optimization, they can be classified into one category, and only one of them can be employed as an "independent objective" attribute to participate in the evaluation.

Especially, if more "objectives" than "independent objective" participate in the analysis and evaluation of multi-objective optimization problem, for example, both flexural strength and impact toughness of such group of tool materials participate in the evaluation simultaneously, it is equivalent to increase their weighting factors of the relevant objectives. In addition, there are some methods for clustering analysis related to fuzzy theory as well [11–17].

Furthermore, as to the problem of selection of cutting tool material, the preferences are with higher hardness, flexural strength, thermal conductivity, and heat-resistant temperature, and lower coefficient of thermal expansion. Therefore, according to the probability-based multi-objective optimization, the attributes of hardness, flexural

Table 9.2 Physical and mechanical properties of eight samples of tool materials

Material	Attribute					
	Hardness (HRA) × 0.1, *A*	Flexural strength (GPa), *B*	Impact toughness (kJ/m^2) × 0.1, *C*	Thermal conductivity (W/m·K), *D*	Coefficient of thermal expansion (10^{-6}/K), *E*	Heat-resistant temperature (K), *F*
W18 Cr4V	8.31	3.2	25	20.9	11	893
YG6	8.95	1.45	3	79.6	4.5	1173
YG6x	9.1	1.4	2	79.6	4.4	1173
YG8	8.9	1.5	4	75.4	4.5	1173
YT30	9.25	0.9	0.3	20.9	7.0	1273
YT14	9.05	1.2	0.7	33.5	6.21	1173
Al_2O_3 AM	9.1	0.5	0.5	19.2	7.9	1473
Si_3N_4 SM	9.2	0.8	0.4	38.2	1.75	1573

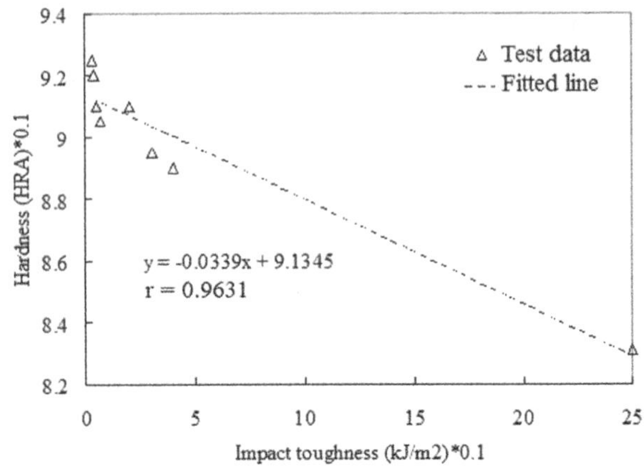

Fig. 9.2 Correlation of hardness vs impact toughness of tool materials

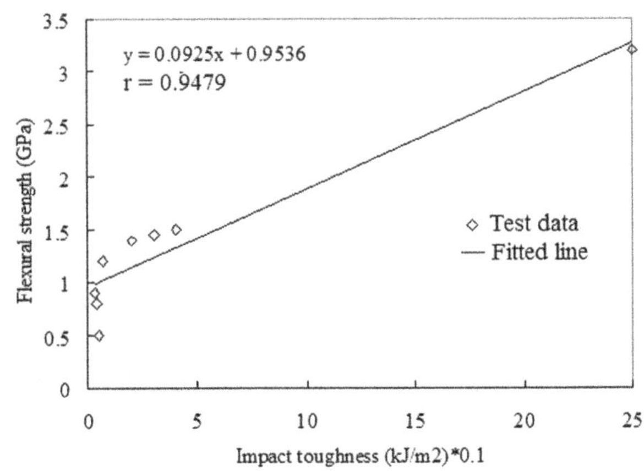

Fig. 9.3 Correlation of flexural strength vs impact toughness of tool materials

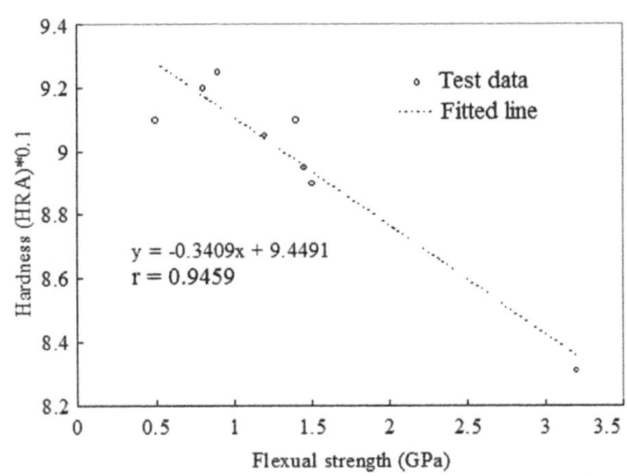

Fig. 9.4 Correlation of hardness versus flexural strength of tool materials

Table 9.3 Assessed consequences of the eight tool materials

Material	Partial preferable probability					Total preferable probability	Rank
	P_A	P_B	P_D	P_E	P_F	$P_t \times 10^5$	
W18Cr4V	0.1156	0.2922	0.0569	0.0320	0.0902	0.1965	8
YG6	0.1245	0.1324	0.2167	0.1507	0.1184	8.1186	2
YG6 x	0.1266	0.1279	0.2167	0.1525	0.1184	8.1644	1
YG8	0.1239	0.1370	0.2053	0.1507	0.1184	7.9110	3
YT 30	0.1287	0.0822	0.0569	0.1050	0.1285	0.6643	6
YT 14	0.1259	0.1096	0.0912	0.1195	0.1184	1.7968	5
Al_2O_3 ceramic	0.1266	0.0457	0.0523	0.0886	0.1487	0.2373	7
Si_3N_4 ceramic	0.1280	0.0731	0.1040	0.2010	0.1588	3.9282	4

Notice: YG indicates hard alloy with the main contents of W and Co; YT indicates hard alloy with the main contents of W, Ti, and Co.

strength, thermal conductivity, and heat-resistant temperature belong to the beneficial type of performance index to participate in the evaluation of partial probability, and coefficient of thermal expansion belongs to the unbeneficial type of performance index to participate in the evaluation of partial probability.

The assessed consequences of preferable probability and ranking are shown in Table 9.3. It indicates that the total preferable probability, the top three are YG6x, YG6, and YG8, which are headed by YG6x.

9.4 Conclusion

Using linear correlation coefficient analysis method, the separation of an independent objective (attribute) from multiple objectives can be conducted properly. In the evaluation, if the attributes of "non-independent objectives" are involved instead of only "independent objectives" participating in the analysis and evaluation of multi-objective optimization simultaneously, it is equivalent to increase their weighting factors of the relevant objectives.

References

1. M. Zheng, H. Teng, J. Yu, Y. Cui, Y. Wang, *Probability-Based Multi-Objective Optimization for Material Selection* (Springer, Singapore, 2022)
2. M. Han, *Applied Multivariate Statistical Analysis*, 2nd edn. (Tongji University Press, Shanghai, 2017)
3. Z. Xiao, J. Yu, *Multivariate Statistics and SAS Application*, 2nd edn. (Wuhan University Press, Wuhan, 2013)

4. M. Zheng, J. Yu, in *Proceedings of 8th Virtual International Conference on Science, Technology and Management in Energy, 16–17, Dec., 2022, Belgrade Serbia*, ed. by L.Z. Velimirović. Cluster Analysis in Probability—Based Multi—Objective Optimization, (Math. Inst. of the Serbian Academy of Sci. & Arts, Belgrade, Serbia, 2023), pp. 313–317.
5. X. Yan, Evaluation model of agricultural machinery selection based on similarity coefficient and distance. Anhui Agric. Sci. **39**(19), 11888–11891 (2011)
6. D. Xu, M. Xia, X. Liu, Index synthesis method of similarity coefficient and its application. J. Huazhong Normal Univ. **29**(2), 155–157 (1995)
7. B. Xu, Rationality of the definition of similarity coefficient. Eng. Math. **8**(3), 13–20 (1992)
8. M. Zhong, Similarity coefficient and distance between component data. J. Math. Med. **9**(2), 145–147 (1996)
9. H. Yan, S. Li, Y. Huang, X. Zhang, Test of common similarity criteria and use of comprehensive similarity coefficient. Meteorol. Technol. **31**(4), 211–215 (2003)
10. S. Cao, Fuzzy cluster analysis of physical and mechanical properties of cutting tool materials. J. Tianjin Textile Inst. Technol. **16**(5), 26–30 (1997)
11. S. Yu, J. Zhang, J. Li, Determination method of mechanical properties of materials based on fuzzy clustering. J. Gansu Sci. **13**(1), 6–11 (2001)
12. R. Shao, Application of fuzzy cluster analysis method in system analysis. Liaoning Chem. Indus. **31**(9), 386–389 (2002)
13. C. Yang, D. Hong, Multi-objective decision method based on fuzzy cluster analysis. J. Xi'an Univ. Finance Econ. **18**(4), 31–34 (2005)
14. J. Gan, Fuzzy cluster analysis of cutting performance of carbide tools. J. Harbin Inst. Shipbuilding Eng. **10**(1), 40–53 (1989)
15. P. Giordani, M.B. Ferrao, F. Martella, *An Introduction to Clustering with R* (Springer, Singapore, 2020)
16. W. Shitong, K.F. Chung, S. Hongbin, Z. Ruiqiang, Note on the relationship between probabilistic and fuzzy clustering. Soft. Comput. **8**, 366–369 (2004)
17. W. Pedrycz, *An Introduction to Computing with Fuzzy Sets* (Springer, Cham, 2021)

Chapter 10
Applications of Probability-Based Multi-objective Optimization Beyond Material Selection

Abstract The probability-based multi-objective optimization could be widely applied in other fields as well; here in this chapter, the application of multi-objective optimization in drug design and extraction is given in details first. It includes the water-soluble chitosan/poly-gamma-glutamic acid-tanshinone IIA with response surface design and glycerosome–triptolide with orthogonal experimental design, compatibility of the traditional Chinese medicine drug by using orthogonal experimental design, and optimal drug extraction conditions based on uniform experimental designs. Other usages, such as scheme selections for military engineering project with weighting factor, and water purification treatment, are demonstrated briefly.

10.1 Introduction

Nowadays, most optimal problems are actually multi-objective optimization problems (MOO). The main characteristics of multi-objective optimal problems are the contradiction and non-commutability between objectives; that is, the improvement of one objective performance may induce decreases in other objective performance. Especially, there is no general metric between the objectives, and therefore, they cannot be compared to each other directly [1]. The optimized solution of a multi-objective problem is actually a set of possible solutions, which is usually called the non-inferior solution set, such as the famous Pareto solution set. Many problems concerning optimization in drug study and preparation are multi-objective optimization ones. As an example, it needs to adjust the compatibility of the drug efficacy and the side effect of a drug. In the evaluation, the balance in a scheme design should be conducted by adjusting some points so that the therapeutic effective target is at a better level and the side effect is relatively minimal. Such kind of problems of multi-objective evaluation occurred usually in preparation of encapsulation of drug with biopolymer in setting drug loading efficiency and encapsulation efficiency as optimal objectives [2]. Besides, in the exploration study of Chinese herbal drugs, the dose–effect relationship of Chinese herbals has nonlinear features, and there exists big difference in the efficacy of different doses of prescriptions, and the efficacy of Chinese herbal drugs has specific multiple routines, points, and multiple targets.

M. Zheng et al., *Probability-Based Multi-objective Optimization for Material Selection*, https://doi.org/10.1007/978-981-99-3939-8_10

The research of traditional Chinese medicine compound drugs involves a number of different drug combinations, and the difference of dose of each drug may lead to different drug effects. Therefore, it is necessary to find out the drugs that play a major role, the effect of drug compatibility and the optimal ratio of doses. The workload will be quite large if the screening is conducted by doing entire test for each drug and dose through approach of one by one. In order to gain an efficient and scientific result of the test, the test could be designed by means of experimental design methodology. The appropriate evaluation of the effect of a drug or compatibility of a compound drug involves the comprehensive consideration of these effect indicators with the help of multi-objective optimization, which will be beneficial to obtain the optimal parameters in line with clinical and experimental reality [3–6].

Application of multi-objective optimization method to multi-objective optimization in preparation of encapsulation composite with designed test and drug extraction conditions are given in the following sections in detail first. It includes the water-soluble chitosan/poly-gamma-glutamic acid-tanshinone IIA with response surface design and glycerosome–triptolide with orthogonal experimental design, compatibility of the traditional Chinese medicine drug by using orthogonal experiment design, and optimal drug extraction conditions based on uniform experimental designs.

Other applications, such as scheme selections for military engineering project with weighting factor, and water purification treatment, are demonstrated briefly.

10.2 Application of the Multi-objective Optimization in Drug Design and Extraction

10.2.1 Optimal Preparation of Encapsulation Composite of Water-Soluble Chitosan/Poly-Gamma-Glutamic Acid-Tanshinone IIA with Response Surface Methodology Design

Optimal preparation of water-soluble chitosan/poly-gamma-glutamic acid-tanshinone IIA as an encapsulation composite was evaluated by Yu et al. with response surface methodology [2]. The traditional treatment of response surface design is with "additive" algorithm to deal with the utilities of multiple objectives frequently. As was pointed in [1], the inherent shortcomings of subjective and artificial factors in "additive" algorithm exist in the traditional treatment for multi-objective optimizations [1]. So, the retreatment by using probability-based multi-objective optimization is valuable.

Table 10.1 collected the analysis results of the utilities in the optimal preparation of water-soluble chitosan (WCS)/poly-gamma-glutamic acid (γ-PGA)-tanshinone IIA (TA) as an encapsulation composite by using response surface methodology

[2], which is renamed as *WCS-γ-PGA-TA* in short hereafter. The independent input variables contain x_1, x_2, x_3, and x_4. The meaning of x_1 is the concentration of WCS (mg ml^{-1}), x_2 is concentration of TA (mg ml^{-1}), x_3 presents the ratio of TA to carrier material (in weight), and x_4 reflects reaction time (h). The drug loading efficiency and encapsulation efficiency are the objectives, which are attributed to the beneficial type index. Table 10.2 gives the fundamental factors and level of the Box–Behnken test design.

Table 10.1 Schemes of Box–Behnken experiment design and results for WCS-γ-PGA-TA

Scheme	x_1	x_2	x_3	x_4	Drug loading efficiency Y_l (%)	Encapsulation efficiency Y_e (%)
1	1	0	1	1	5.25	79.31
2	0	0	0	0	11.22	93.25
3	0	0	0	0	9.92	94.31
4	0	0	− 1	1	6.38	85.22
5	0	1	1	0	4.38	72.51
6	0	0	− 1	− 1	5.18	75.87
7	1	0	0	− 1	6.97	84.56
8	0	0	0	0	10.09	90.34
9	0	− 1	0	1	6.38	85.69
10	1	1	0	0	6.39	79.84
11	0	0	1	1	7.89	87.21
12	− 1	0	0	− 1	8.73	92.80
13	1	0	0	1	6.34	89.96
14	0	0	1	− 1	6.05	79.65
15	0	− 1	0	− 1	5.08	80.79
16	− 1	1	0	0	3.96	66.73
17	0	1	0	− 1	4.26	78.62
18	− 1	0	1	0	6.32	76.97
19	− 1	0	0	1	9.58	90.73
20	0	− 1	1	0	4.73	78.22
21	1	0	− 1	0	6.21	78.34
22	1	0	0	0	5.07	84.97
23	− 1	− 1	0	0	6.01	84.46
24	0	− 1	0	0	6.32	83.36
25	0	0	0	1	11.03	95.02
26	0	− 1	− 1	0	4.98	80.33
27	0	1	− 1	0	5.38	70.67
28	0	0	0	0	9.89	92.73
29	− 1	0	− 1	0	6.54	80.39

Table 10.2 Fundamental factors and levels of the Box–Behnken test design

Fundamental factor	Level		
	−1	0	1
x_1: WCS concentration (mg ml^{-1})	5	6	7
x_2: TA concentration (mg ml^{-1})	0.5	1.0	1.5
x_3: Ratio of TA to carrier material (in weight)	1:3	1:5	1:7
x_4: Reaction time (h)	0.5	1.0	1.5

Table 10.3 presents the evaluated consequences for preferable probability of the utility in the preparation of *WCS-γ-PGA-TA* by using the new probability-based multi-objective optimization method with response surface design. From Table 10.3, it can be seen that the tests *Nos.* 25 and 2 are the proper schemes for the preparation of *WCS-γ-PGA-TA* by using response surface design in the first glance comparatively.

Further optimization could be gained by regressions for the data in Table 10.2.

Equation (10.1) is the regressed result for the total preferable probability P_t *vs* independent input variables, x_1, x_2, x_3, and x_4.

$$\begin{aligned} P_t \times 10^3 = {} & 1.9644 - 0.1787x_1 + 0.0254x_2 \\ & - 7.2 \times 10^{-5}x_3 + 0.1717x_4 - 0.4411x_1^2 \\ & - 0.7445x_2^2 - 0.5602x_3^2 - 0.1494x_4^2 \\ & + 0.3381\,x_1x_3 - 0.0329x_1x_4 - 0.0172x_2x_3 \\ & + 0.0913x_2x_4 + 0.0443x_3x_4 \\ R^2 = {} & 0.8620. \end{aligned} \tag{10.1}$$

The total preferable probability P_t gains its maximum value $P_{t\max} \times 10^3 = 2.0394$ at $x_1 = 5.755$ mg ml^{-1}, $x_2 = 1.0275$ mg ml^{-1}, $x_3 = 1{:}4.9$, and $x_4 = 1.302$ h.

Similarly, drug loading efficiency Y_l (%) and encapsulation efficiency Y_e (%) of preparation are regressed as well, which are given as follows:

$$\begin{aligned} Y_l(\%) = {} & 10.0612 - 0.8559x_1 + 0.2232x_2 \\ & - 0.0247x_3 + 0.7516x_4 - 1.9370x_1^2 \\ & - 3.3194x_2^2 - 2.3077x_3^2 - 0.7432x_4^2 \\ & + 1.5714x_1x_3 - 0.2467x_1x_4 - 0.1875x_2x_3 \\ & + 0.4711x_2x_4 + 0.2385x_3x_4 \\ R^2 = {} & 0.8420. \end{aligned} \tag{10.2}$$

Y_l obtains its optimal value $Y_{l\,\text{Opt.}} = 10.40\%$ at $x_1 = 5.755$ mg ml^{-1}, $x_2 = 1.0275$ mg ml^{-1}, $x_3 = 1{:}4.9$, and $x_4 = 1.302$ h.

Analogically,

Table 10.3 Evaluated consequences of preferable probability of utility in the preparation of WCS-γ-PGA-TA

Scheme	Partial preferable probability		Total preferable probability
	P_l	P_e	$P_t \times 10^3$
1	0.0267	0.0329	0.8781
2	0.0571	0.0387	2.2064
3	0.0505	0.0391	1.9729
4	0.0325	0.0353	1.1466
5	0.0223	0.0301	0.6698
6	0.0264	0.0314	0.8288
7	0.0355	0.0350	1.2429
8	0.0513	0.0374	1.9223
9	0.0325	0.0355	1.1529
10	0.0325	0.0331	1.0759
11	0.0401	0.0361	1.4511
12	0.0444	0.0385	1.7085
13	0.0323	0.0373	1.2028
14	0.0308	0.0330	1.0162
15	0.0258	0.0335	0.8655
16	0.0202	0.0277	0.5573
17	0.0217	0.0326	0.7063
18	0.0322	0.0319	1.0258
19	0.0487	0.0376	1.8330
20	0.0241	0.0324	0.7802
21	0.0316	0.0325	1.0259
22	0.0258	0.0352	0.9085
23	0.0306	0.0350	1.0705
24	0.0322	0.0345	1.1110
25	0.0561	0.0394	2.2102
26	0.0253	0.0333	0.8436
27	0.0274	0.0293	0.8018
28	0.0503	0.0384	1.9340
29	0.0333	0.0333	1.1087

$$\begin{aligned} Y_e = {} & 92.4514 - 0.8660x_1 - 2.6375x_2 \\ & + 0.1389x_3 + 2.2449x_4 - 4.8949x_1^2 \\ & - 10.3491x_2^2 - 8.7025x_3^2 - 0.1760x_4^2 \\ & + 6.1970x_1x_3 + 0.7516x_1x_4 + 0.9875x_2x_3 \\ & - 0.6621x_2x_4 - 0.6855x_3x_4 \end{aligned}$$

$$R^2 = 0.9060. \tag{10.3}$$

Y_e gains its optimal value $Y_{e\,\mathrm{Opt.}} = 93.43\%$ at $x_1 = 5.755$ mg ml^{-1}, $x_2 = 1.0275$ mg ml^{-1}, $x_3 = 1{:}4.9$, and $x_4 = 1.302$ h.

The optimal consequences for $Y_{e\,\mathrm{Opt.}}$ and $Y_{l\,\mathrm{Opt.}}$ are close to the tested values of drug loading and encapsulation efficiency, i.e., 10.29% and 91.89%, respectively.

10.2.2 Optimal Preparation of Glycerosomes–Triptolide as an Encapsulation Composite with Orthogonal Experimental Design

Optimal design of glycerosome formulations to enhance transdermal triptolide delivery was conducted by Zhu et al. with orthogonal experimental design [6], the drug loading Y_l (%) and entrapment efficiency Y_e (%) of the nanocarriers are employed as objectives for the optimization. The independent input variables include glycerol concentration A (%), phospholipid to cholesterol mass ratio B (m/m) and phospholipid to triptolide mass ratio C (m/m); three-level orthogonal table L_9 (3^4) was employed, Table 10.4 cited the arrangement of experiment design and consequences on basis of orthogonal design L_9 (3^4) [6].

Again, the encapsulation efficiency Y_e and drug loading efficiency Y_l are attributed to the beneficial type indexes. Table 10.5 represents the assessed consequences of preferable probability of the experimental data while Table 10.6 shows the assessed results of range analysis for total preferable probability. The optimal configuration is $A_2B_3C_1$, which is fortunately the same as the first glanced maximum of test scheme *No.* 6 in Tables 10.4 and 10.5.

Table 10.4 Experimental arrangement and results based on the L_9 (3^4) orthogonal design

Scheme	A (%)	B (m/m)	C (m/m)	Y_l (%)	Y_e (%)
1	1	1	1	15.41	65.67
2	1	2	2	5.97	61.87
3	1	3	3	3.12	55.79
4	2	1	2	5.71	65.56
5	2	2	3	3.07	54.64
6	2	3	1	16.19	77.40
7	3	1	3	2.93	43.25
8	3	2	1	15.97	67.37
9	3	3	2	6.06	54.85

Table 10.5 Assessed results of preferable probability of this experimental data

Scheme	P_{Yl}	P_{Ye}	$P_t \times 10^2$	Rank
1	0.2070	0.1202	2.4883	3
2	0.0802	0.1132	0.9082	5
3	0.0419	0.1021	0.4280	7
4	0.0767	0.1200	0.9205	4
5	0.0412	0.1000	0.4125	8
6	0.2175	0.1417	3.0813	1
7	0.0394	0.0792	0.3116	9
8	0.2146	0.1233	2.6455	2
9	0.0814	0.1004	0.8173	6

Table 10.6 Assessed results of range analysis for total preferable probability of glycerosome formulations

Level	*A*	*B*	*C*
1	1.2749	1.2401	2.7384
2	1.4714	1.3221	0.8820
3	1.2581	1.4422	0.3840
Range	0.2133	0.2021	2.3544
Order	2	3	1

10.2.3 Optimization of Compatibility of the Traditional Chinese Medicine Drug by Using Orthogonal Experimental Design

Compatibility of the traditional Chinese medicine drug for Poria, Guizhi, Atractylodes, and Licorice compound medicine was conducted by Song et al. with orthogonal experiment design and correlation analysis [5].

In the orthogonal experiment design, there are *four* independent variables, i.e., contents of Poria labeled *A*, Guizhi labeled *B*, Atractylodes labeled *C*, and Licorice labeled *D*. The independent input variables are designed with *four* levels while the diuretic labeled Y_1, anti-hypoxia labeled Y_2, and anti-ventricular fibrillation effect labeled Y_3 are taken as the evaluation objectives.

Diuretic and anti-hypoxia have the characteristic of the larger the better, and thus they are attributed to the beneficial type indexes, and the anti-ventricular fibrillation effect is measured by the rate of ventricular fibrillation, which has characteristic of the smaller the better, and belongs to the unbeneficial type index. Table 10.7 presents the arrangement of the orthogonal design and the tested results. Table 10.8 displays the assessed results of orthogonal experiment design.

Table 10.8 indicates that the scheme *No.* 16 possesses the highest value of the total preferable probability, which can be chosen as the proper scheme in the first glance.

Table 10.7 Arrangement of orthogonal test and test results

Scheme	A (g)	B (g)	C (g)	D (g)	Y_1 (g)	Y_2 (min)	Y_3
1	0	0	0	0	1.1559	20.75	64.3411
2	0	4.5	4.5	3	1.2999	23.56	56.0122
3	0	9	9	6	1.3525	26.46	48.5904
4	0	13.5	13.5	9	1.4019	30.31	48.5904
5	6	0	4.5	6	1.2371	23.24	56.0122
6	6	4.5	0	9	1.4157	24.57	52.2388
7	6	9	13.5	0	1.6832	29.35	45.0000
8	6	13.5	9	3	1.8287	31.28	45.0000
9	12	0	9	9	1.3239	26.67	48.5904
10	12	4.5	13.5	6	1.5575	31.23	45.0000
11	12	9	0	3	1.7384	31.65	37.7612
12	12	13.5	4.5	0	1.8174	32.15	41.4096
13	18	0	13.5	3	1.4874	31.53	48.5904
14	18	4.5	9	0	1.6901	30.87	41.4096
15	18	9	4.5	9	1.7769	31.56	37.7612
16	18	13.5	0	6	1.8545	32.28	33.9878

Table 10.8 Assessed results of the orthogonal experiment design and results by using the probability-based multi-objective optimization

Scheme	Preferable probability			Total	
	Y_1	Y_2	Y_3	$P_t \times 10^4$	Rank
1	0.0469	0.0454	0.0413	0.8795	16
2	0.0528	0.0515	0.0514	1.3982	14
3	0.0549	0.0578	0.0604	1.9203	11
4	0.0569	0.0663	0.0604	2.2801	10
5	0.0502	0.0508	0.0514	1.3125	15
6	0.0575	0.0537	0.0560	1.7296	13
7	0.0684	0.0642	0.0648	2.8423	7
8	0.0743	0.0684	0.0648	3.2910	5
9	0.0538	0.0583	0.0604	1.8947	12
10	0.0633	0.0683	0.0648	2.7985	8
11	0.0706	0.0692	0.0736	3.5952	3
12	0.0738	0.0703	0.0692	3.5880	4
13	0.0604	0.0689	0.0604	2.5165	9
14	0.0686	0.0675	0.0692	3.2038	6
15	0.0722	0.0670	0.0736	3.6644	2
16	0.0753	0.0706	0.0782	4.1554	1

Table 10.9 Assessed results of range analysis for total preferable probability of drug compatibility

Level	Poria *A*	Guizhi *B*	Atractylodes *C*	Licorice *D*
*k*1	1.6195	1.6508	2.5899	2.6284
*k*2	2.2939	2.2825	2.4908	2.7002
*k*3	2.9691	3.0055	2.5775	2.5467
*k*4	3.3850	3.3286	2.6093	2.3922
Range	1.7655	1.6778	0.1186	0.3080
Order	1	2	4	3

Furthermore, range analysis is conducted for the total preferable probability; the corresponding results are shown in Table 10.9. Table 10.9 represents that the optimal configuration is $A_4B_4C_4D_2$, i.e., "*Poria* 18 g, *Guizhi* 13.5 g, *Atractylodes* 13.5 g, and *Licorice* 3 g."

10.2.4 Optimization of Multi-objective Drug Extraction Conditions Based on Uniform Experimental Designs

Optimization of multi-objective drug extraction conditions was conducted by Wu et al. by using uniform experimental design and NSGA to gain a Pareto non-inferior solution [4]. Analogically, the inherent shortcomings of Pareto non-inferior solution are addressed in the previous section due to its "additive" algorithm and scaling factors [1].

In the uniform experiment design [4], the independent input variables contain microwave power x_1 (W), ethanol concentration x_2 (%), extraction time x_3 (min), ethanol consumption x_4 (times), and pulverization degree x_5 (mesh). The evaluated objectives include extract rate Y_1 (%), schisandrin A Y_2 (%), and total lignans Y_3 (%) [4].

Table 10.10 represents the results of the microwave extraction of Schisandrin in uniform experimental design $U_{10}(10^8)$. The performance indexes extract rate Y_1, schisandrin A Y_2, and total lignans Y_3 all belong to beneficial type indexes. Table 10.11 displays the evaluated consequences of this problem with uniform experimental design $U_{10}(10^8)$ by using probability-based multi-objective optimization. Table 10.11 indicates that the scheme *No.* 6 is with the highest value of the total preferable probability, which can be chosen as the proper scheme in the first glance. Besides, the regression analysis for the total preferable probability P_t with respect to the independent input variables, x_1, x_2, x_3, x_4, and x_5 can be done to gain further optimization.

$$P_t \times 10^3 = -6.0063 + 0.0029x_1 + 0.1730x_2 - 0.0132x_3 - 0.1994x_4 + 0.0260x_5 - 2.8 \times 10^{-6}x_1^2$$

$$- 0.0011x_2^2 + 0.0002x_3^2 + 0.0147x_4^2,$$
$$R^2 = 1. \qquad (10.4)$$

The total preferable probability P_t gains its maximum value $P_{tmax} \times 10^3 = 3.8385$ at $x_1 = 526.79$ W, $x_2 = 75.89\%$, $x_3 = 5$ min, $x_4 = 12$, and $x_5 = 80$ mesh.

Analogically, the regressed extract rate Y_1 (%), schisandrin A Y_2 (%), and total lignans Y_3 (%) are given as follows:

$$Y_1(\%) = 112.0776 - 0.0231x_1 - 1.6962x_2 - 0.4366x_3 - 4.5526x_4$$

Table 10.10 Uniform experiment design $U_{10}(10^8)$ and results for microwave extraction of Schisandrin

Scheme	x_1 (W)	x_2 (%)	x_3 (min)	x_4 (times)	x_5 (mesh)	Y_1 (%)	Y_2 (%)	Y_3 (%)
1	170	60	15	8	80	27.08	3.71	8.08
2	170	70	35	12	60	28.19	3.64	7.92
3	340	90	5	6	40	19.99	3.25	8.39
4	340	50	25	12	20	35.51	1.41	3.96
5	510	60	45	6	0	17.56	0.59	3.64
6	510	80	5	10	80	25.41	4.39	10.42
7	680	90	25	4	60	18.62	4.08	9.77
8	680	50	45	10	40	30.39	1.78	5.58
9	850	70	15	4	20	23.42	2.01	6.75
10	850	80	35	8	0	15.54	0.51	3.47

Table 10.11 Assessed consequences for the microwave extraction of Schisandrin in uniform test design $U_{10}(10^8)$

Scheme	Preferable probability			Total	
	Y_1	Y_2	Y_3	$P_t \times 10^3$	Rank
1	0.1120	0.1462	0.1189	1.9473	3
2	0.1166	0.1435	0.1165	1.9495	2
3	0.0827	0.1281	0.1234	1.3076	5
4	0.1469	0.0556	0.0583	0.4756	8
5	0.0726	0.0233	0.0535	0.0905	9
6	0.1051	0.1730	0.1533	2.7883	1
7	0.0770	0.1608	0.1437	1.7805	4
8	0.1257	0.0702	0.0821	0.7241	7
9	0.0970	0.0792	0.0993	0.7622	6
10	0.0643	0.0201	0.0510	0.0660	10

$$+ 0.0489x_5 - 2.49 \times 10^{-5}x_1^2 + 0.0100x_2^2$$
$$+ 0.0056x_3^2 + 0.3413x_4^2,$$
$$R^2 = 1 \tag{10.5}$$

Y_1 get its optimal value $Y_{1\text{opt.}} = 31.77\%$ at $x_1 = 526.79$ W, $x_2 = 75.89\%$, $x_3 = 5$ min, $x_4 = 12$, and $x_5 = 80$ mesh.

$$\begin{aligned} Y_2(\%) = &\ 3.0058 - 0.0008x_1 + 0.0166x_2 \\ &- 0.0436x_3 - 0.7514x_4 + 0.0405x_5 \\ &+ 2.85 \times 10^{-7}x_1^2 + 5.59 \times 10^{-5}x_2^2 \\ &+ 0.0008x_3^2 + 0.0450x_4^2. \end{aligned}$$
$$R^2 = 1. \tag{10.6}$$

Y_2 obtains its optimal value $Y_{2\text{opt.}} = 5.04\%$ at $x_1 = 526.79$ W, $x_2 = 75.89\%$, $x_3 = 5$ min, $x_4 = 12$, and $x_5 = 80$ mesh.

$$\begin{aligned} Y_3(\%) = &\ 7.5496 - 0.0002x_1 \\ &+ 0.0781x_2 - 0.1758x_3 - 1.6514x_4 \\ &+ 0.0703x_5 + 7.32 \times 10^{-7}x_1^2 - 0.0002x_3^2 \\ &+ 0.0032x_3^2 + 0.0961x_4^2. \end{aligned}$$
$$R^2 = 1. \tag{10.7}$$

Y_3 get its optimal $Y_{3\text{opt.}} = 11.18\%$ at $x_1 = 526.79$ W, $x_2 = 75.89\%$, $x_3 = 5$ min, $x_4 = 12$, and $x_5 = 80$ mesh.

Obviously, the optimal results from regression are superior to the chosen in the first glance.

10.3 Application of the Probability-Based Multi-objective Optimization in Military Engineering Project with Weighting Factor

10.3.1 Decision Making of Multi-objective Military Engineering Investment

Zhou conducted a study of decision making of military engineering investment with utility function [7]. A field facility project is to be evaluated. After preliminary feasibility study, four options are preselected as candidates with *three* decision objectives, i.e., investment C_1, period C_2, and efficiency C_3. The relevant data is shown

Table 10.12 Relevant data for military engineering investment project

Option	Investment, C_1 (Ten thousand ¥ RMB)	Period, C_2 (months)	Efficiency, C_3 (%)
A_1	4800	9	78
A_2	5200	11	75
A_3	3900	12	82
A_4	4300	14	90
Weighting factor, w	0.228	0.575	0.137

Table 10.13 Preferable probabilities and ranking of the military engineering investment project

Option	Partial preferable probability			Total	
	P_{C1}	P_{C2}	P_{C3}	P_t	Rank
A_1	0.2363	0.3043	0.2400	0.2986	1
A_2	0.2143	0.2609	0.2308	0.2659	3
A_3	0.2857	0.2391	0.2523	0.2734	2
A_4	0.2637	0.1957	0.2769	0.2422	4

in Table 10.12. In the objectives, efficiency C_3 belongs to beneficial type of performance indicator, investment C_1 and period C_2 are attributed to unbeneficial type of performance indicators.

The assessed preferable probabilities and ranking are shown in Table 10.13, which shows that the Option *No.* A_1 is the proper scheme.

As the weighting factor is involved, the total preferable probability is assessed by,

$$P_i = P_{i1}^{w_1} \cdot P_{i2}^{w_2} \cdots P_{im}^{w_m} = \prod_{j=1}^{m} P_{ij}^{w_j} \tag{10.8}$$

In Eq. (10.8), w_j is the weighting factor of jth indicator in the evaluation.

10.3.2 Flexible Ability Assessment of Antiaircraft Weapon System

Feng et al. performed the flexible ability assessment of antiaircraft weapon system based on combination weighting for MADM [8]. Table 10.14 cites the utilities of the evaluation indexes U_1, U_2, U_3, and U_4 of the schemes S_1, S_2, S_3, and S_4 with the weighting w [8]. All utilities are attributed to beneficial type of performance indicators. The assessed preferable probabilities and ranking are given in Table 10.15, which shows that the scheme *No.* S_3 is the proper choice.

Table 10.14 Utilities of the evaluation indexes for the schemes

Scheme	Index of utility			
	U_1	U_2	U_3	U_4
S_1	8.54	5.48	4.00	6.22
S_2	9.42	4.20	7.26	5.83
S_3	8.70	8.40	7.50	6.50
S_4	7.42	7.25	8.45	5.10
Weighting factor, w	0.383	0.267	0.150	0.200

Table 10.15 Assessed preferable probabilities and ranking of the flexible ability for antiaircraft weapon system

Scheme	Partial preferable probability				Total	
	P_{U1}	P_{U2}	P_{U3}	P_{U4}	P_t	Rank
S_1	0.2506	0.2163	0.1470	0.2630	0.2246	4
S_2	0.2764	0.1658	0.2668	0.2465	0.2344	3
S_3	0.2553	0.3316	0.2756	0.2748	0.2810	1
S_4	0.2177	0.2862	0.3105	0.2156	0.2465	2

10.4 Comparative Analysis of Scheme Selection for Water Purification Treatment by Using PMOO with the Traditional MCDM

Robule Lake was influenced by waste materials which contain mining activities. Eight schemes for purification treatment of the water from Robule Lake were comparatively analyzed by Štirbanović et al. by using the traditional multi-criteria decision makings (MCDM) [9]. The contaminate materials contain various metal ions, such as Fe, Zn, Cu, Mn, Ni, and Cd, their acid mine drainage (AMD) treatment methods include, passive treatment method S_1, sequential neutralization S_2, ion exchange S_3, adsorption process with low-cost adsorbents S_4, adsorption process with natural zeolites S_5, electrodialysis S_6, filtration with nanofiltration membranes S_7, and last one reverse osmosis S_8. Seven evaluation objectives (criteria) were employed for the scheme selection of the water purification treatment methods, which contain, removal efficiency of the metal ions and quality of the purified water E_1, investment costs E_2, possibility of reuse of the treated waste E_3, desirability of post-treatment and/or pretreatment E_4, maintenance and operation costs E_5, sensitivity of scheme E_6, field needed E_7. There is weighting factor in their treatment [9]. Five MCDM methods were employed in their evaluations, which include TOPSIS, MOOSRA, VIKOR, CoCoSo, and WASPAS. The utility for each evaluation index per scheme is cited and presented in Table 10.16. The meanings of "max" and "min" in Table 10.16 stand for the beneficial and unbeneficial types of the utility indexes. Table 10.17 shows the evaluated consequences by using PMOO method. Table 10.18 cites the

Table 10.16 Initial decision-making matrix of the treatment scheme

Objective	E_1	E_2	E_3	E_4	E_5	E_6	E_7
Optimization	max	min	max	max	max	max	max
S_1	6	3	5	10	10	8	2
S_2	9	7	10	10	9	9	9
S_3	9	3	1	3	3	6	8
S_4	5	7	1	10	5	6	8
S_5	5	7	1	10	5	1	3
S_6	9	3	1	3	3	1	3
S_7	9	3	1	3	3	1	3
S_8	9	3	1	3	3	1	3
Weighting factor	0.30	0.20	0.10	0.20	0.10	0.05	0.05

Table 10.17 Evaluated consequences by using PMOO method

No.	Partial preferable probability							Total	
	P_{E1}	P_{E2}	P_{E3}	P_{E4}	P_{E5}	P_{E6}	P_{E7}	P_t	Rank
S_1	0.0984	0.1591	0.2381	0.1923	0.2439	0.2424	0.0513	0.1802	2
S_2	0.1475	0.0682	0.4762	0.1923	0.2195	0.2727	0.2308	0.1975	1
S_3	0.1475	0.1591	0.0476	0.0577	0.0732	0.1818	0.2051	0.1275	3
S_4	0.0820	0.0682	0.0476	0.1923	0.1220	0.1818	0.2051	0.1208	4
S_5	0.0820	0.0682	0.0476	0.1923	0.1220	0.0303	0.0769	0.1052	8
S_6	0.1475	0.1591	0.0476	0.0577	0.0732	0.0303	0.0769	0.1110	5
S_7	0.1475	0.1591	0.0476	0.0577	0.0732	0.0303	0.0769	0.1110	5
S_8	0.1475	0.1591	0.0476	0.0577	0.0732	0.0303	0.0769	0.1110	5

evaluation results by using traditional MCDM methods together with our assessed result of using PMOO for comparison [9].

From Table 10.18, it can be seen that the result of ranking of WASPAS is the same as that of PMOO by chance, while most of others are different except for S_2.

10.5 Application of the Probability-Based Multi-objective Optimization in Power Equipment

Ofodu et al. applied the probability-based multi-objective optimization in high voltage thermofluids to optimize the composition [10]. It showed that the most promising candidate for high-voltage equipment is the nanofluid with 0.6 wt% Al_2O_3 comprehensively. It also identified the inadequacy of using the produced Jatropha oil for high voltage equipment. Abifarin et al. used the probability-based multi-objective

Table 10.18 Comparison of the evaluation results by using traditional MCDM methods with PMOO [9]

No.	Rank					
	TOPSIS	MOOSRA	VIKOR	CoCoSo	WASPAS	PMOO
S_1	2	4	3	2	2	2
S_2	1	1	1	1	1	1
S_3	3	5	2	3	3	3
S_4	7	2	7	7	4	4
S_5	8	3	8	8	8	8
S_6	4	6	4	4	5	5
S_7	4	6	4	4	5	5
S_8	4	6	4	4	5	5

optimization to determine the best oil from a lot of data of the produced oils for power equipment application [11]. The results showed that a mixture of Jatropha oil and neem oil with adding nanoparticle was the best for power equipment. It showed also that the probability-based multi-objective optimization was successfully employed [11].

10.6 Conclusion

By using the newly developed probability-based multi-objective optimization method, the applications beyond material selection are given in detail. The results indicate the applicability of the methodology.

References

1. M. Zheng, Y. Wang, H. Teng, A new "intersection" method for multi-objective optimization in material selection. Tehnicki Glas. **15**(4), 562–568 (2021). https://doi.org/10.31803/tg-20210901142449
2. J. Yu, N. Wu, X. Zheng, M. Zheng, Preparation of water—soluble chitosan/poly-gama-glutamic acid—tanshinone IIA encapsulation composite and its in vitro/in vivo drug release properties. Biomed. Phys. Eng. Express **6**, 045020 (2020)
3. M. Chen, X. Lu, Q. Zhu, L. Xu, Evaluation of common chemotherapy regimens in advanced non-small cell lung adenocarcinoma based on multi–attribute utility theory. Chin. J. Drug Appl. Monit. **18**(1), 1–4 (2021)
4. X. Wu, F. Li, C. Liu, Multi–objective optimize based on nondominated sorting genetic algorithm—uniform multi-objective optimization of extraction conditions of drug application. Chin. J. Health Stat. **30**(2), 177–181 (2013)

5. Z. Song, D. Feng, J. Xu, K. Bi, Study on the compatibility and therapeutical basis of composite herbal medicines of Lingguishugan decoction. Chin. Traditional Patent Med. **25**(2), 132–137 (2003)
6. C. Zhu, Y. Zhang, T. Wu, Z. He, T. Guo, N. Feng, Optimizing glycerosome formulations via an orthogonal experimental design to enhance transdermal triptolide delivery. Acta Pharm. **72**(1), 135–146 (2022)
7. S.F. Zhou, Multi-attribution effectiveness policy decision of military project investment. Mil. Econ. Res. **27**(8), 54–57 (2006)
8. H. Feng, H. Mao, F. Zeng, C. Zhang, Flexible ability assessment of antiaircraft weapon system based on combination weighting for MADM. Mod. Defense Technol. **42**(5), 13–18 (2014)
9. Z. Štirbanović, V. Gardić, D. Stanujkić, R. Marković, J. Soković, Z. Stevanović, Comparative MCDM analysis for AMD treatment method selection. Water Resour. Manag. **35**, 3737–3753 (2021). https://doi.org/10.1007/s11269-021-02914-3
10. J.C. Ofodu, J.K. Abifarin, Employment of probability—based multi-response optimization in high voltage thermofluids. Mil. Tech. Courier **70**(2), 393–408 (2022). https://doi.org/10.5937/vojtehg70-35764;doi:10.5937/vojtehg70-35764
11. J.K. Abifarin, J.C. Ofodu, Determination of an efficient power equipment oil through a multi–criteria decision making analysis. Mil. Tech. Courier **70**(2), 433–446 (2022). https://doi.org/10.5937/vojtehg70-36024;doi:10.5937/vojtehg70-36024

Chapter 11
Treatment of Portfolio Investment by Means of Probability-Based Multi-objective Optimization

Abstract In this chapter, the portfolio investment problem is treated by using probability-based multi-objective optimization together with "uniform design for experiments with mixtures." It involves the simultaneous optimization of maximum return rate and minimum risk of the portfolio investment as optimal problem with double objectives. The optimal problem with double objectives of maximum return rate and minimum risk is transferred into a mono-objective by means of probability-based multi-objective optimization first, the "uniform design for experiments with mixtures" is used to perform discretization for the succeeding data treatment. The analysis reveals that the probability-based multi-objective optimization methodology could result in a rational optimal consequence of the related problems.

11.1 Introduction

Investment is a specific economic activity in modern society. Investors usually expect to receive highest returns with least risk for their limited funds. In fact, the return of an investment is always interwoven with its risk. Actually, a high return is often accompanied by a high risk in general.

In order to reduce or even avoid big risks, funds can be often diversified into some securities. On a number of securities, the existence of many securities in the market provides them with guarantee, which is due to the effect of so-called portfolio investment. Early in 1952, Harry M. Markowitz put forward a theory of "Portfolio Selection," which opens a prelude of modern portfolio research, since then it provides him the honor of founder of the theory of contemporary securities portfolio. The emergence of Markowitz's portfolio investment theory has promoted the innovation of financial market theory, which is considered as an innovation of financial theory. It initiates the beginning of modern portfolio investment theory [1].

The original version of this chapter was revised: Equation 11.8 has been corrected. The correction to this chapter is available at
https://doi.org/10.1007/978-981-99-3939-8_15

M. Zheng et al., *Probability-Based Multi-objective Optimization for Material Selection*, https://doi.org/10.1007/978-981-99-3939-8_11

In Markowitz's theory, there are two assessing indexes of risky securities, namely, the average investment return rate ρ and the square variance of return rate s^2 or the standard deviation of return rate s; the former (ρ) shows a measurement of the profitability index of securities, while the latter (s^2 or s) reflects a measurement of risk index of securities [2]. Furthermore, the optimal portfolio could be determined by one of the following two models.

Model A:

$$\text{Min } s^2 = X^T CX \tag{11.1}$$

$$\text{s.t. } \rho = \sum_{j=1}^{m} x_j \rho_j \geq \alpha,$$

$$\sum_{j=1}^{m} x_j = 1.$$

Model B:

$$\text{Max } \rho = \sum_{j=1}^{m} x_j \rho_j \tag{11.2}$$

$$\text{s.t. } s^2 = X^T CX \leq \beta,$$

$$\sum_{j=1}^{m} x_j = 1.$$

In Eqs. (11.1) and (11.2), the number m reflects the number of securities, and the expected yields of each security are represented by $\rho_1, \rho_2, \ldots, \rho_m$, respectively; the proportional ratio of the jth security is x_j. C reflects the risk matrix of investment, and $X^T CX$ reflects the expected value of portfolio risk. Moreover, $s^2 = X^T CX$ is used to represent the variance of the return rate of securities portfolio, which is used to reflect the investment risk.

A function $f_1 = E(R) = \sum_{j=1}^{m} x_j \rho_j$ is utilized to express the expected value of the return rate of m kinds of securities invested in a certain period of time. Factor α reflects the preset total return of portfolio investment. Factor β describes the preset risk of portfolio investment. s^2 indicates the deviation of various possible values of return rate from their expected values, that is, the uncertainty of return rate. The standard deviation of portfolio f_2 is the square root of variance s^2.

Obviously, the portfolio investment theory of Markowitz adopted either limiting the risk to a certain range to obtain as much profit as possible, or limiting the profit to a certain range to suffer as little risk as possible. This is similar to the ε–constraint solution in the of multi-objective optimization problem. Its obvious disadvantage is that "one objective optimization" is used instead of "multi-objective simultaneous optimization," which thus losses the intrinsic essence of "simultaneous optimization of multiple objectives."

Recently, the probability-based multi-objective optimization was proposed [3], which is with a brand new concept of "preferable probability" and the assessments for the probability-based multi-objective optimization. As a rationally novel approach, it aims to conduct the overall optimization of the system in spirit of probability theory. The methodology could be used in many fields, including energy planning, programming problem, operation research, financial affairs, etc. Here in this chapter, the portfolio investment problem is retreated by using probability-based multi-objective optimization, and furthermore the "uniform design for experiments with mixtures" is used to perform discretization for the succeeding data treatment.

11.2 Solution of Portfolio Problem by Means of Probability-Based Multi-objective Optimization

In light of Markowitz's theory, a return rate function f_1 and a risk function f_2 are introduced as, respectively,

$$f_1 = E(R) = \sum_{j=1}^{m} x_j \rho_j, \tag{11.3}$$

$$\begin{aligned} f_2 = {} & [(x_1 s_1)^2 + (x_2 s_2)^2 + (x_3 s_3)^2 + \cdots + (x_n s_n)^2 + \eta_{1,2}(x_1 s_1)(x_2 s_2) \\ & + \eta_{1,3}(x_1 s_1)(x_3 s_3) + \eta_{1,4}(x_1 s_1)(x_4 s_4) + \cdots + \eta_{i,j}(x_i s_i)(x_j s_j) \\ & + \cdots + \eta_{m-1,m}(x_{m-1} s_{m-1})(x_m s_m)]^{0.5}. \end{aligned} \tag{11.4}$$

In Eq. (11.4), $\eta_{i,j}$ is the correlation coefficient between the ith security and the jth security; s_j is the risk of jth security.

According to the assessment method of the probability-based multi-objective optimization methodology for objective [3], f_1 belongs to the beneficial type of utility index and f_2 belongs to the unbeneficial type of utility index.

Therefore, the "portfolio investment" problem is a typical optimal problem with double objectives, i.e., the simultaneous optimization of maximum return rate and minimum risk of the portfolio investment. Therefore, the probability-based multi-objective optimization methodology can be utilized reasonably.

11.3 Example of Case with Four Securities

The combination of four securities is taken as an example, i.e., $m = 4$ in Eqs. (11.3) and (11.4). The specific optimization process is explained in details.

Assume the expected return rate of security ***A*** is $\rho_1 = 10\%$, and the standard deviation of return is $s_1 = 3\%$. The expected return rate of ***B*** securities $\rho_2 = 15\%$, and the standard deviation of return rate is $s_2 = 5\%$. The expected return rate of ***C*** securities is $\rho_3 = 18\%$, and the standard deviation of return rate is $s_3 = 6\%$. The expected return rate of ***D*** securities is $\rho_4 = 7\%$, and the standard deviation of return rate is $s_4 = 2\%$. Furthermore, it is assumed following correlation coefficients between two of above securities, $\eta_{1,2} = 0.84$, $\eta_{1,3} = 0.81$, $\eta_{1,4} = -0.58$, $\eta_{2,3} = 0.64$, $\eta_{2,4} = -0.42$, $\eta_{3,4} = -0.49$.

Now it needs to seek the proper ratio of this portfolio investment, such that it obtains simultaneous optimization of maximum return rate and minimum risk.

Solution

In this section, the problem of "portfolio" is retreated with the help of probability-based multi-objective optimization methodology.

In this optimization problem of double objectives, set the investment percentages of four securities, ***A***, ***B***, ***C***, and ***D***, as x_1, x_2, x_3, and x_4, respectively. While it has a constraint condition of $x_1 + x_2 + x_3 + x_4 = 1$, therefore it has actually only three independent variables, namely x_1, x_2, and x_3.

In principle, the assessment of the simultaneous optimization of maximizing return rate and minimizing risk of portfolio problem can be formulated as,

$$\text{Max} \quad f_1 = x_1\rho_1 + x_2\rho_2 + x_3\rho_3 + x_4\rho_4 \tag{11.5}$$

$$\begin{aligned} \text{Min} \ \ f_2 = {} & [(x_1s_1)^2 + (x_2s_2)^2 + (x_3s_3)^2 + (x_4s_4)^2 + \eta_{1,2}(x_1s_1)(x_2s_2) \\ & + \eta_{1,3}(x_1s_1)(x_3s_3) + \eta_{1,4}(x_1s_1)(x_4s_4) + \eta_{2,3}(x_2s_2)(x_3s_3) \\ & + \eta_{2,4}(x_2s_2)(x_4s_4) + \eta_{3,4}(x_3s_3)(x_4s_4)]^{0.5} \end{aligned} \tag{11.6}$$

$$\text{s.t.:} \ \sum_{j=1}^{4} x_j = 1. \tag{11.7}$$

The partial preferable probability P_{f1} of f_1 is assessed according to beneficial type of objective index of the probability-based multi-objective optimization methodology, and the assessment of partial preferable probability P_{f2} of f_2 is conducted in accordance with the unbeneficial one [3], i.e.,

$$\begin{aligned} & P_{f1} = \lambda \times f_1(x_1, x_2, x_3, x_4), \\ & \kappa = 1/\left[\int_{x_1, x_2, x_3 = 0}^{1, \text{s.t.} \sum_{j=1}^{4} x_j = 1} f_1(x_1, x_2, x_3, x_4) \mathrm{d}x_1 \, \mathrm{d}x_2 \, \mathrm{d}x_3\right], \end{aligned} \tag{11.8}$$

$$P_{f_2} = \lambda \times [f_{2\max}(x_1, x_2, x_3, x_4) + f_{2\min}(x_1, x_2, x_3, x_4) - f_2(x_1, x_2, x_3, x_4)],$$

$$\lambda = 1/\left\{ \int_{x_1,x_2,x_3=0}^{1,\text{s.t. } \sum_{j=1}^{4} x_j = 1} [f_{2\max}(x_1, x_2, x_3, x_4) + f_{2\min}(x_1, x_2, x_3, x_4) - f_2(x_1, x_2, x_3, x_4)] \mathrm{d}x_1\, \mathrm{d}x_2\, \mathrm{d}x_3 \right\}. \quad (11.9)$$

Furthermore, the total preferable probability P_t is,

$$P_t = P_{f_1} \times P_{f_2}. \quad (11.10)$$

The maximum of P_t corresponds to the optimum point of simultaneous optimization of maximizing return rate and minimizing risk of this portfolio problem.

Obviously, the assessments of Eqs. (11.8) and (11.9) are rather complex actually.

In order to simplify the solving of Eqs. (11.8) and (11.9), discretization treatment is introduced.

Fang once proposed a "uniform design for experiments with mixtures" to perform the discretization of experimental design with the constraint of $\sum_{j=1}^{4} x_j = 1$ [4, 5]. Now, this approach is adopted to conduct our discretization.

Since the constraint condition of $x_1 + x_2 + x_3 + x_4 = 1$, there are only three independent variables in fact, says x_1, x_2, and x_3, while the sampling points are positioned in four-dimensional space indeed; therefore, it needs to include at least 23 sampling points with characteristic of "good lattice point" in the effective region for the discretization of processing according to our previous study [6–8]. Thus, according to Fang [4, 5], the uniform table $U_{23}^*(23^7)$ can be taken as the initial table to construct a uniform test design table $UM_{23}(23^4)$ with mixtures, as shown in Table 11.1. The concrete steps are as follows:

(I) **Choice of the uniform design table**

In general, as to a given number of mixtures m ($m = 4$ here in our problem) and the number of sampling points n (23 in our problem), the corresponding table $U_n^*(n^t)$ or $U_n(n^t)$ and the usage table from the uniform design table can be chosen, which were shown by Fang [4, 5], and the number of columns of the usage table is selected as $m - 1$ under such condition. Furthermore, the original elements in the uniform design table $U_n^*(n^t)$ or $U_n(n^t)$ are marked by $\{q_{ik}\}$.

(II) **Build a new element c_{ki}**

For each i, build its new element c_{ki} by the following formula, $c_{ki} = (2q_{ki} - 1)/(2n)$.

(III) **Build uniform sampling points for the mixtures, x_{ki}**

It employs the following formula to conduct the matter,

$$x_{ki} = (1 - c_{ki}^{\frac{1}{m-i}}) \prod_{j=1}^{i-1} c_{kj}^{\frac{1}{m-j}}, \quad i = 1, \ldots, m - 1$$

Table 11.1 Uniform test table $UM_{23}(23^4)$ with mixtures based on uniform design table $U^*_{23}(23^7)$

No.	q_{10}	q_{20}	q_{30}	c_1	c_2	c_3	x_1	x_2	x_3	x_4
1	**11**	**17**	**19**	**0.4565**	**0.7174**	**0.8044**	**0.2300**	**0.1178**	**0.1276**	**0.5246**
2	22	10	14	0.9348	0.4130	0.5870	0.0222	0.3494	0.2596	0.3688
3	9	3	9	0.3696	0.1087	0.3696	0.2824	0.4810	0.1492	0.0874
4	20	20	4	0.8478	0.8478	0.1522	0.0535	0.0750	0.7389	0.1326
5	7	13	23	0.2826	0.5435	0.9783	0.3438	0.1725	0.0105	0.4733
6	18	6	18	0.7609	0.2391	0.7609	0.0871	0.4665	0.1068	0.3397
7	5	23	13	0.1957	0.9783	0.5435	0.4195	0.0063	0.2621	0.3121
8	16	16	8	0.6739	0.6739	0.3261	0.1233	0.1570	0.4850	0.2347
9	3	9	3	0.1087	0.3696	0.1087	0.5228	0.1871	0.2586	0.0315
10	14	2	22	0.5870	0.0652	0.9348	0.1627	0.6235	0.0139	0.1999
11	1	19	17	0.0217	0.8043	0.7174	0.7209	0.0288	0.0707	0.1796
12	12	12	12	0.5	0.5	0.5	0.2063	0.2325	0.2806	0.2806
13	23	5	7	0.9783	0.1957	0.2826	0.0073	0.5536	0.3150	0.1241
14	10	22	2	0.4130	0.9348	0.0652	0.2553	0.0247	0.6731	0.0470
15	21	15	21	0.8913	0.6304	0.8913	0.0376	0.1982	0.0831	0.6811
16	8	8	16	0.3261	0.3261	0.6739	0.3117	0.2953	0.1282	0.2649
17	19	1	11	0.8043	0.0217	0.4565	0.0700	0.7929	0.0745	0.0626
18	6	18	6	0.2391	0.7609	0.2391	0.3793	0.0793	0.4119	0.1295
19	17	11	1	0.7174	0.4565	0.0217	0.1048	0.2903	0.5917	0.0131
20	4	4	20	0.1522	0.1522	0.8478	0.4661	0.3256	0.0317	0.1766
21	**15**	**21**	**15**	**0.6304**	**0.8913**	**0.6304**	**0.1425**	**0.0479**	**0.2992**	**0.5104**
22	2	14	10	0.0652	0.5870	0.4130	0.5975	0.0941	0.1810	0.1274
23	13	7	5	0.5435	0.2826	0.1957	0.1839	0.3822	0.3490	0.0849

$$x_{km} = \prod_{j=1}^{m-1} c_{kj}^{\frac{1}{m-j}}, \quad k = 1, \ldots, n.$$

Thus, $\{x_{ik}\}$ is used to set up the corresponding uniform design table $UM_n(n^m)$ of the mixture under the conditions of m and n.

Uniform test table $UM_{23}(23^4)$ with mixtures of Table 11.1 is based on uniform design table $U^*_{23}(23^7)$. Because here $m = 4$, $n = 23$, from above rules, $x_{k1} = 1-c_{k1}^{1/3}$, $x_{k2} = c_{k1}^{1/3} \times \left(1-c_{k2}^{1/2}\right)$, $x_{k3} = c_{k1}^{1/3} \times c_{k2}^{1/2} \times (1 - c_{k3})$, $x_{k4} = c_{k1}^{1/3} \times c_{k2}^{1/2} \times c_{k3}$.

Further, we can get values of the yield function f_1 and risk function f_2, and the distribution of their preferable probability and ranking at the sampling points, as shown in Table 11.2.

Figure 11.1 shows the positions of sampling points and optimum points. The results show that the maximum total preferable probability appears at the 21st discrete

Table 11.2 Evaluation results of yield f_1, risk f_2, preferable probability and ranking at sampling points

No.	Values of f_1 and f_2		Partial preferable probability		Overall preferable probability	Rank
	f_1	f_2	P_{f1}	P_{f2}	$P_t \times 10^3$	
1	**0.1004**	**0.0161**	**0.0345**	**0.0656**	**2.2621**	**2**
2	0.1272	0.0279	0.0437	0.0493	2.1537	5
3	0.1334	0.0375	0.0458	0.0359	1.6453	16
4	0.1589	0.0471	0.0546	0.0227	1.2386	21
5	0.0953	0.0158	0.0327	0.0660	2.1593	4
6	0.1217	0.0277	0.0418	0.0495	2.0692	7
7	0.1119	0.0246	0.0385	0.0539	2.0719	6
8	0.1396	0.0360	0.0480	0.0381	1.8258	13
9	0.1291	0.0371	0.0444	0.0365	1.6189	17
10	0.1263	0.0343	0.0434	0.0403	1.7510	14
11	0.1017	0.0245	0.0350	0.0539	1.8838	10
12	0.1257	0.0292	0.0432	0.0474	2.0470	8
13	0.1492	0.0415	0.0513	0.0305	1.5613	19
14	0.1537	0.0472	0.0528	0.0224	1.1853	22
15	0.0961	0.0146	0.0330	0.0677	2.2359	3
16	0.1171	0.0270	0.0402	0.0505	2.0312	9
17	0.1437	0.0439	0.0494	0.0270	1.3348	20
18	0.1330	0.0364	0.0457	0.0375	1.7143	15
19	0.1615	0.0489	0.0555	0.0202	1.1215	23
20	0.1135	0.0288	0.0390	0.0480	1.8711	11
21	**0.1110**	**0.0205**	**0.0382**	**0.0594**	**2.2676**	**1**
22	0.1154	0.0301	0.0397	0.0462	1.8330	12
23	0.1445	0.0406	0.0497	0.0317	1.5735	18

sampling point tightly followed by the 1st discrete sampling point, and therefore, they can be used as the optimal solution of this portfolio problem.

As to the 21st sampling point, the investment ratio is $x_1^* = 0.1425$, $x_2^* = 0.0479$, $x_3^* = 0.2292$, $x_4^* = 0.5204$, and the obtained rate of return is 11.10% and the risk is 2.05%.

As to the 1st sampling point, its investment ratio is, $x_1^* = 0.2300$, $x_2^* = 0.1178$, $x_3^* = 0.1276$, $x_4^* = 0.5246$, and the obtained rate of return is 10.04% and the risk is 1.61%.

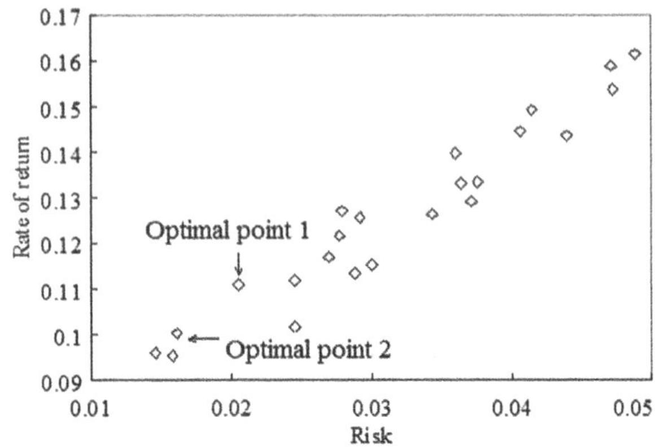

Fig. 11.1 Location of discrete sampling points and optimization points

11.4 Conclusion

In this chapter, the probability-based multi-objective optimization method is employed to deal with the simultaneous optimization of portfolio investment problem with maximum return rate and minimum risk of investment. The analysis shows that the probability-based multi-objective optimization methodology could result in the optimal solution of related problems. The "uniform design for experiments with mixtures" can be rationally used to conduct the discretization for the simplification.

References

1. A. Xi, L. Sun, X. Rong, S. Yang, Studies of optimization portfolio models. J. Tianjin Inst. Textile Sci. Technol. **19**(3), 8–10 (2000)
2. R. Hu, Comparative study on various models of portfolio. Oper. Res. Manag. Sci. **10**(1), 98–103 (2001)
3. M. Zheng, H. Teng, J. Yu, Y. Cui, Y. Wang, *Probability-Based Multi-objective Optimization for Material Selection* (Springer, Singapore, 2022). https://doi.org/10.1007/978-981-19-3351-6
4. K.-T. Fang, *Uniform Design and Uniform Design Table* (Science Press, Beijing, 1994). https://doi.org/10.1007/7-03-004290-5/O743
5. K.-T. Fang, M. Liu, H. Qin, Y. Zhou, *Theory and Application of Uniform Experimental Design* (Springer/Science Press, Beijing/Singapore, 2018). https://doi.org/10.1007/978-981-13-2041-5
6. J. Yu, M. Zheng, Y. Wang, H. Teng, An efficient approach for calculating a definite integral with about a dozen of sampling points. Vojnotehnički Glas. Mil. Techn. Courier **70**(2), 340–356 (2022). https://doi.org/10.5937/vojtehg70-36029
7. M. Zheng, H. Teng, Y. Wang, J. Yu, Appropriate algorithm for assessment of numerical integration, in *Proceedings of 2022 International Joint Conference on Information and Communication*, 20–22 May (Seoul, 2022), pp. 18–22. https://doi.org/10.1109/cice56791.2022.00015
8. M. Zheng, J. Yu, H. Teng, Y. Wang, Application of probability-based multi-objective optimization in portfolio investment and engineering management problems. Tehnički Glas. **18**(2) (2024). https://doi.org/10.31803/tg-20221202130111

Chapter 12
Treatment of Multi-objective Shortest Path Problem by Means of Probability-Based Multi-objective Optimization

Abstract An actual transportation process is with multiple objectives generally. However, the previous approaches could not derive the rational shortest path with optimizing all multiple objectives at the same time appropriately. In this chapter, the multi-objective shortest path problem of transportation process is treated by means of probabilistic methodology to obtain a rational solution. Each objective is analogically taken as an "individual event," the simultaneous optimization of multiple objectives is equivalent of the "joint event" of simultaneous occurrence of the "multiple events"; thus the simultaneous optimization of multiple objectives can be conducted analogically by means of probabilistic methodology. The partial preferable probability of each objective of every routine (scheme) is evaluated according to the actual preference degree of utility index of the objective. Moreover, the product of all partial preferable probabilities of utility indexes of objective of each routine (scheme) forms the total preferable probability of the corresponding routine (scheme), which indicates the uniquely decisive index of the routine (scheme) in the multi-objective shortest path problem in spirit of probability theory. The optimal solution of the multi-objective shortest path problem is the routine (scheme) with the highest total preferable probability. Finally, two application examples are given to illuminate the approach.

12.1 Introduction

In network optimization, the shortest problem (SP) can be seen as a classic problem. The shortest problem with single objective has been well solved by Dijkstra's algorithm and Floyd's algorithm [1, 2]. However, in 1990s some new problems raised since the development of information science, modern communication, and intelligent network. These new problems are basically the shortest path problem with multiple objectives, which makes the study of the shortest path problem active again [3–8].

Usually, the minimization of an objective, such as cost and transportation time, is often considered for general shortest path problems. However, it is often necessary to consider multiple objectives in transportation network in the route selection at the

M. Zheng et al., *Probability-Based Multi-objective Optimization for Material Selection*, https://doi.org/10.1007/978-981-99-3939-8_12

same time as well, such objective as cost, time, risk, safety, etc., different objectives have to be compromised in the solution process simultaneously.

The classification and generalization of the research status of multi-objective shortest path problem were stated by Current et al. [9]. There are three types of methods for solving multi-objective shortest path problem mainly: (1) Utility function method, which requires prior preference information of the decision maker to determine the corresponding utility function; (2) Interactive method, which uses preference information in the entire problem-solving process; (3) Production method, which solves the complete Pareto optimization or approximate optimization solution set directly, it mainly includes dynamic programming method, Pareto marking method, and Pareto rank method.

For the multi-objective shortest path algorithm, the usual processing rule is to linearly weight different objectives with "additive algorithm" or convert some objectives into constraints. But for the linear weighting method, the determination of its weight is very problematic. Besides, the "additive algorithm" implies the union operation in respect of set theory. For the constrained shortest path problem, it has been proved to be NP-hard. In fact, above algorithms not only deviate from the intrinsic intention of multi-objective actually, but also consume large time and space, or even unsolved when the scale of the problem is large [10, 11].

The dual-objective shortest path problem is a common situation in the multi-objective shortest path problem. In order to solve the dual-objective shortest path problem, Current et al. and Coutinaho-Rodrigues et al. studied the general dual-objective shortest path algorithm and proposed an interactive dual-objective algorithm [12, 13]. In the shortest path problem with dual objectives, it is often necessary to obtain an effective path and then select it. Hansen and Climaco et al. got certain achievement on the acquisition of effective paths for dual objectives [14, 15].

In this chaper, a probability-based shortest path approach is proposed to conduct shortest path problem with optimizing multiple objectives at the same time. It takes each objective as an "individual event" analogically first, and thus the simultaneous optimization of multiple objectives is equivalent to the "joint event" of concurrent occurrence of the "multiple events." The partial preferable probability of each objective of every routine (scheme) and the total preferable probability of each routine (scheme) are evaluated as the uniquely decisive indexes of the routine (scheme) in the multi-objective shortest path problem by means of probability theory. Examples are given to illuminate the approach as well.

12.2 Approach for Multi-objective Shortest Path Problem Based on Probability Theory

12.2.1 Probabilistic Model of Multi-objective Optimization Problem

According to the recently proposed probability-based multi-objective optimization (PMOO) for material selection [16, 17], each objective can be analogically taken as an "individual event," and simultaneous optimization of multiple objectives is thus the equivalent of the "joint event" of concurrent occurrence of "multiple events" in respect of probability theory formally. Thus, the problem of simultaneous optimization of multiple objectives is converted into a joint probabilistic problem of simultaneous occurrence of the "multiple events." Furthermore, the preference degree of utility index of each objective of a routine (scheme) is transferred into a partial preferable probability of the corresponding "event," the total/overall preferable probability of the "integral event" (simultaneous optimization of multiple objectives) is rationally the product of all the partial preferable probabilities of all corresponding events of the routine (scheme) in the respect of probability theory. In the evaluation, the utility of the objective is preliminarily classified as beneficial or unbeneficial type according to the characteristic of the preference or function in the assessment.

12.2.2 Assessment Procedure of Simultaneous Optimization of Multi-objective Shortest Path Problem in Respect of Probability Theory

(1) **Assessment of Objective and Event**

As to the multi-objective shortest path problem, the cost, time, risk, safety, etc., are different objectives, of which each objective can be taken as "individual event" analogically. Thus, the multi-objective shortest path problem is transferred into the joint probabilistic problem of simultaneous occurrence of "multiple events" equivalently. The accumulated cost, time, risk, safety, etc., in each route (scheme) are accounted for according to the actual interval of each route (scheme) individually.

(2) **Assessment of Preferable Probability**

The partial preferable probabilities of cost, time, risk, safety, etc., are assessed according to their specific function or characteristic of the preference for every scheme analogically. Finally, the overall/total preferable probability of the "integral event" is the product of all the partial preferable probabilities of all events of each route (scheme) in the spirit of probability theory, which thus completes the simultaneous optimization of multi-objective and indicates the uniquely decisive index of the route (scheme) in the multi-objective shortest path problem in respect of probability theory.

(3) **Solution of the Multi-objective Shortest Path Problem**

At last, the optimal solution of the multi-objective shortest path problem is the route (scheme), which is with the highest total preferable probability.

12.3 Application of the Probability-Based Approach of Multi-objective Shortest Path Problem

12.3.1 Application in Hazardous Materials Transportation Path Problem

The hazardous materials transportation path problem is an important thing, the cost, risk, population coverage, and transit time are taken as evaluation objectives [18].

The transportation network is shown in Fig. 12.1, and now the problem is to obtain the shortest path in the transportation of hazardous goods [18]. As can be seen from Fig. 12.1, there are five possible transportation options (route/scheme) from the start point ***S*** to destination ***T***.

(1) **Without considering weight factor**

Table 12.1 gives the transportation cost, risk, population coverage, and transit time between various nodes in the transportation process without considering weight factor. Table 12.2 shows the accounted values of the five transportation schemes to the above four objective indicators when starting from point ***S*** to destination ***T***, which is accounted from the data in Table 12.1 and Fig. 12.1.

Fig. 12.1 Transportation network

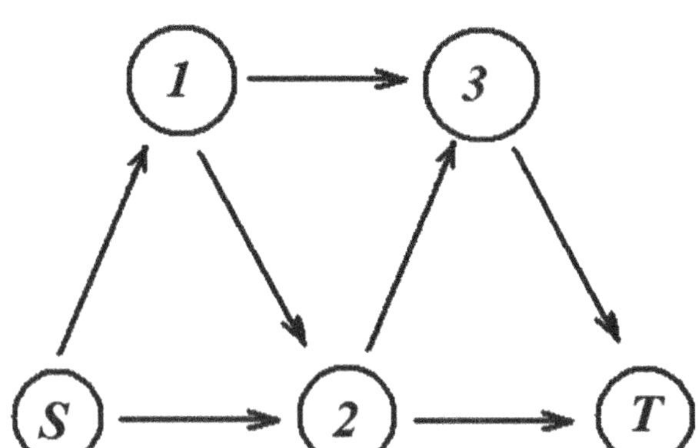

Table 12.1 Transportation cost, risk, population coverage, and transit time for each interval

Interval	Transportation cost (¥RMB)	Risk	Population coverage (person)	Transit time (h)
$(S \rightarrow 1)$	40	20	30	4
$(S \rightarrow 2)$	20	50	100	8
$(1 \rightarrow 2)$	60	10	110	4
$(1 \rightarrow 3)$	30	10	20	3
$(2 \rightarrow 3)$	80	10	120	6
$(2 \rightarrow T)$	50	25	60	3
$(3 \rightarrow T)$	40	20	120	5

Table 12.2 Accounted values of the possible transport schemes from ***S*** to ***T***

Scheme	Transportation cost, *C* (¥RMB)	Risk, *R*	Population coverage, *P* (person)	Transit time, *T* (h)
$A(S \rightarrow 1 \rightarrow 2 \rightarrow 3 \rightarrow T)$	220	60	380	19
B(S → 1 → 3 → T)	**110**	**50**	**170**	**12**
$C(S \rightarrow 1 \rightarrow 2 \rightarrow T)$	150	55	200	11
$D(S \rightarrow 2 \rightarrow T)$	70	75	160	11
$E(S \rightarrow 2 \rightarrow 3 \rightarrow T)$	140	80	340	19

The objectives, i.e., cost, risk, population coverage, and transit time in this transportation problem, are all unbeneficial type of indexes for each scheme in the assessment. The evaluations of partial preferable probabilities and the total preferable probabilities P_i for each of the five possible transport options can be conducted according to the procedure of the probability-based approach of multi-objective.

Table 12.3 presents the partial preferable probabilities and the total preferable probabilities P_i for each of the five possible transport options from ***S*** to ***T***. Table 12.3 shows that scheme ***B*(*S* → 1 →3 → *T*)** exhibits the highest total preferable probability, which might be selected as the optimum route.

Table 12.3 Assessments of partial preferable probabilities and the total preferable probabilities for each scheme from ***S*** to ***T***

Scheme	P_C	P_R	P_P	P_T	$P_i \times 10^3$	Rank
$A(S \rightarrow 1 \rightarrow 2 \rightarrow 3 \rightarrow T)$	0.0921	0.2113	0.1103	0.1410	0.3028	5
B(S → 1 → 3 → T)	**0.2368**	**0.2394**	**0.2552**	**0.2308**	**3.3393**	**1**
$C(S \rightarrow 1 \rightarrow 2 \rightarrow T)$	0.1842	0.2254	0.2345	0.2436	2.3711	3
$D(S \rightarrow 2 \rightarrow T)$	0.2895	0.1690	0.2621	0.2436	3.1232	2
$E(S \rightarrow 2 \rightarrow 3 \rightarrow T)$	0.1974	0.1549	0.1379	0.1410	0.5948	4

Table 12.4 Assessments of partial preferable probabilities and the total preferable probabilities of each scheme from ***S*** to ***T*** with weight factors (weighted 0.1, 0.4, 0.3, 0.2)

Scheme	P_C	P_R	P_P	P_T	P_i	Rank
$A(S \to 1 \to 2 \to 3 \to T)$	0.0921	0.2113	0.1103	0.1410	0.1476	5
B(S → 1 → 3 → T)	**0.2368**	**0.2394**	**0.2552**	**0.2308**	**0.2420**	**1**
$C(S \to 1 \to 2 \to T)$	0.1842	0.2254	0.2345	0.2436	0.2270	2
$D(S \to 2 \to T)$	0.2895	0.1690	0.2621	0.2436	0.2189	3
$E(S \to 2 \to 3 \to T)$	0.1974	0.1549	0.1379	0.1410	0.1541	4

(2) **Considering Weight Factor**

If there is weight factor for the objective, the weight factor can be taken as the exponent of the corresponding partial preferable probability in the product of individual partial preferable probability for the total preferable probability assessment [16, 17]. Here, for this problem, let's assume the weight factors for the transportation cost, risk, population coverage, and transit time is 0.1, 0.4, 0.3, and 0.2, respectively. Thus, the evaluation can be conducted accordingly. Table 12.4 gives the partial preferable probabilities and the total preferable probabilities of each scheme from ***S*** to ***T*** with the weight factors of 0.1, 0.4, 0.3, and 0.2, individually.

Table 12.4 shows that scheme ***B*(*S* → 1 → 3 → *T*)** displays the highest total preferable probability P_i luckily, which might be selected as the optimum route.

12.3.2 Application in Multi-objective Inter-Model Transportation of Grain from Northern China to the South Considering Weather Factor

Multi-objective inter-model transportation of grain from northern China to the south considering weather factor is significant issue, the transportation cost, quality loss, and time can be used as objectives [19].

Since it is a long transportation distance, the transportation through sole highway is expensive, and unique water transportation range is limited, thus multi-modal intermodal transportation has become the best way to transport grain across provinces. The multimodal intermodal transportation of grain in containers has a complex transportation process, which requires network traffic planning and model design at the technical level, in order to cover multiple stages and paths of the intermodal network, and determine the operation with the lowest cost.

According to the change curve of grain quality with temperature, it has been shown that the loss rate of grain quality at high temperature is faster than that at normal temperature and low temperature [19]. Therefore, the proper consideration in the grain transportation process is: pause and rest when encountering high temperature during transportation, and proper shelters, etc., so as to keep the container environment at normal temperature and avoid excessive reduction of grain quality at high temperature.

(1) **Without Considering Weather Factor**

Assuming that a box of corn is transported from Yingkou port to Zhoushan port at 12:00 in summer, 11 potential routines can be considered [19], as shown in Table 12.5.

The objectives, i.e., cost, quality loss, and time in transportation, are all unbeneficial type of indexes of each scheme in the assessment.

Table 12.6 shows the partial preferable probabilities and the total preferable probabilities P_i for each of the 11 potential transport options from Yingkou to Zhoushan. Table 12.6 presents that scheme 7 gives the highest total preferable probability, which might be selected as the optimum route. Besides, scheme 7 is closely followed by scheme 2, the latter might be the backup.

(2) **In Consideration of Weather Factor**

If the weather factor is taken into account, high-temperature pause and rest measures are taken [19], and thus seven routines can be provided, as shown in Table 12.7. It can be seen from Table 12.7 that the transportation routine takes a longer time, while the quality loss is commonly smaller as compared with the case without considering high-temperature pause.

Table 12.8 presents the partial preferable probabilities and the total preferable probabilities P_i for each of the seven possible transport options from Yingkou to Zhoushan.

Table 12.5 Eleven potential routes for corn transportation

No.	Routine
1	Yingkou water–Dalian water–Qingdao water–Lianyungang water–Nantong water–Shanghai water–Zhoushan water
2	Yingkou railway–Dalian Railway–Dalian water–Qingdao water–Lianyungang water–Nantong water–Shanghai water–Zhoushan water
3	Yingkou water–Dalian water–Qingdao water–Lianyungang water–Lianyungang railway–Nantong railway–Shanghai railway–Zhoushan railway
4	Yingkou railway–Dalian railway–Dalian water–Qingdao water–Lianyungang water–Nantong water–Nantong railway–Shanghai railway–Zhoushan railway
5	Yingkou highway–Dalian highway–Dalian water–Qingdao water–Lianyungang water–Nantong water–Shanghai water–Zhoushan water
6	Yingkou water–Dalian water–Qingdao water–Qingdao railway–Lianyungang railway–Nantong railway–Shanghai railway–Zhoushan railway
7	Yingkou railway–Dalian railway–Dalian water–Qingdao water–Lianyungang water–Lianyungang railway–Nantong railway–Shanghai railway–Zhoushan railway
8	Yingkou railway–Dalian railway–Dalian water–Qingdao water–Qingdao railway–Lianyungang railway–Nantong railway–Shanghai railway–Zhoushan railway

(continued)

Table 12.5 (continued)

No.	Routine
9	Yingkou railway–Qinhuangdao railway–Tianjin railway–Jinan railway–Xuzhou railway–Nanjing railway–Nanjing water–Shanghai water–Zhoushan water
10	Yingkou highway–Dalian highway–Dalian water–Qingdao water–Lianyungang water–Lianyungang railway–Nantong railway–Shanghai railway–Zhoushan railway
11	Yingkou railway–Qinhuangdao railway–Tianjin railway–Jinan railway–Xuzhou railway–Nanjing railway–Shanghai railway–Zhoushan railway

No.	Cost, *C* (¥RMB)	Quality loss, *Q* (%)	Time, *T* (h)
1	3002	3.55	66
2	3687	1.59	60
3	4560	5.24	56
4	4685	0.84	64
5	4844	1.575	59
6	5040	3.34	53
7	5245	0.63	50
8	5725	1.38	47
9	6177	3.72	44
10	6402	0.615	49
11	6819	1.245	37

Table 12.6 Preferable probability of optimal route for corn transportation

No.	P_C	P_Q	P_T	$P_i \times 10^3$	Rank
1	0.1315	0.0567	0.0675	0.5032	9
2	**0.1183**	**0.1048**	**0.0785**	**0.9733**	**2**
3	0.1015	0.0151	0.0858	0.1316	11
4	0.0991	0.1233	0.0712	0.8691	4
5	0.0960	0.1052	0.0803	0.8110	6
6	0.0922	0.0618	0.0912	0.5202	8
7	**0.0883**	**0.1284**	**0.0967**	**1.0964**	**1**
8	0.0790	0.1100	0.1022	0.8881	3
9	0.0703	0.0525	0.1077	0.3972	10
10	0.0659	0.1288	0.0985	0.8371	5
11	0.0579	0.1133	0.1204	0.7903	7

Table 12.8 shows that scheme 6 is the highest total preferable probability, which might be selected as the optimum route. Besides, scheme 6 is closely followed by scheme 7, the latter might be the backup.

Table 12.7 Seven possible routines for corn transportation with considering high-temperature pause

No.	Routine
1	Yingkou water–Dalian water–Qingdao water–Lianyungang water–Nantong water–Shanghai water–Zhoushan water
3	Yingkou railway–Dalian railway–Dalian water–Qingdao water–Lianyungang water–Nantong water–Nantong railway–Shanghai railway–Zhoushan railway
3	Yingkou water–Dalian water–Qingdao water–Qingdao railway–Lianyungang railway–Nantong railway–Shanghai railway–Zhoushan railway
4	Yingkou railway–Dalian railway–Dalian water–Qingdao water–Lianyungang water–Lianyungang railway–Nantong railway–Shanghai railway–Zhoushan railway
5	Yingkou railway–Dalian railway–Dalian water–Qingdao water–Qingdao railway–Lianyungang railway–Nantong railway–Shanghai railway–Zhoushan railway
6	Yingkou railway–Qinhuangdao railway–Tianjin railway–Jinan railway–Xuzhou railway–Nanjing railway–Nanjing water–Shanghai water–Zhoushan water
7	Yingkou railway–Qinhuangdao railway–Tianjin railway–Jinan railway–Xuzhou railway–Nanjing railway–Shanghai railway–Zhoushan railway

No.	Cost, C (¥RMB)	Quality loss, Q (%)	Time, T (h)
1	3002	1.23	82
2	3687	1.005	68
3	5040	0.975	65
4	5245	0.855	58
5	5725	0.81	55
6	6177	0.705	48
7	6819	0.6	41

Table 12.8 Preferable probability under high-temperature pause measures

No.	P_C	P_Q	P_T	$P_i \times 10^3$	Rank
1	0.2063	0.0905	0.0923	1.7241	7
2	0.1856	0.1244	0.1239	2.8607	5
3	0.1447	0.1290	0.1306	2.4368	6
4	0.1384	0.1471	0.1464	2.9806	3
5	0.1239	0.1538	0.1532	2.9199	4
6	**0.1103**	**0.1697**	**0.1689**	**3.1601**	**1**
7	**0.0908**	**0.1855**	**0.1847**	**3.1120**	**2**

12.4 Conclusion

From the above discussion, a new approach for multi-objective shortest path problem is developed by means of probability theory. In the assessment, each objective is analogically taken as an "individual event," the preferable probabilities are evaluated according to the preference degree of the corresponding "events" individually. The simultaneous optimization of multiple objectives is equivalent to the simultaneous occurrence of the "multiple events" with their preferable probabilities, which thus completes the simultaneous optimization of multi-objective shortest path problem rationally.

References

1. E.W. Dijkstra, A note on two problems in connection with graphs. Num. Math. **1**, 269–271 (1959)
2. R. Floyd, Algorithm 97: shortest path. Commun. ACM **5**, 345 (1962)
3. X. Cai, T. Kloks, C.K. Wong, Time varying shortest path problems algorithm for problems with constraints. Networks **29**, 141–149 (1997)
4. S.G. Irina Loachim, A dynamic programming algorithm for the shortest path problem with time windows and linear node code. Networks **31**, 193–204 (1998)
5. P. Mirchandani, A simple $O(n^2)$ algorithm for the all - pairs shortest path problem on an interval graph. Networks **27**, 215–217 (1996)
6. D. Burton, L.P. Toint, On an instance of the inverse shortest pairs problem. Math. Program. **53**, 45–61 (1992)
7. G. Yu, J. Yang, On the robust shortest path problem. Comput. Ops. Res. **25**, 457–468 (1998)
8. B. Pelegrim, P. Fernqndez, On the sum-max bi-criterion path problem. Comput. Ops. Res. **25**, 1043–1054 (1998)
9. J. Current, M. Marsh, Multi-objective transportation network design and routing problems: taxonomy and annotation European. J. Oper. Res. **65**, 1–15 (1993)
10. H. Wei, Y. Pu, J. Li, An approach to bi-objective shortest path. Syst. Eng. **23**(7), 113–117 (2005)
11. G. Hao, D. Zhang, D. Wang, A fast algorithm for bi-objective shortest path. J. Highway Transp. Res. Dev. **24**(11), 96–104 (2007)
12. J. Current, C. Revelle, J. Cohon, An interactive approach to identify the best compromise solution for two objective shortest path problems. Comput. Ops. Res. **17**(2), 187–198 (1990)
13. J. Coutinaho-Rodrigues, J. Climcao, J. Current, An interactive bi-objective shortest path approach: search for unsupported non-dominated solutions. Comput. Ops. Res. **26**, 789–798 (1999)
14. P. Hansen, Bicriterion Path Problems, in *Multiple criteria decision making theory and application.* ed. by G. Fandel, T. Gal (Springer, Berlin, Heidelberg, 1979), pp.109–127
15. J.C.N. Clímaco, E.Q.V. Martins, A bicriterion shortest path algorithm. Eur. J. Oper. Res. **11**, 399–404 (1982)
16. M. Zheng, H. Teng, J. Yu, Y. Cui, Y. Wang, *Probability-Based Multi-Objective Optimization for Material Selection* (Springer, Singapore, 2022)
17. M. Zheng, J. Yu, A novel method for solving multi-objective shortest path problem in respect of probability theory. Tehnički Glas. **17**(4), (2023). https://doi.org/10.31803/tg-20220921070537
18. H. Wei, J. Li, J. Wei, An approach for hazardous materials transportation path problem in the time-varying network curfews. J. Industr. Eng. Eng. Manag. **21**(3), 79–85 (2007)
19. C. Feng, H. Zhou, C. Xiang, S. Ni, S. Chen, Research on multi-objective inter-model transportation of grain from northern China to the south considering quality changes. Grain Storage **50**(2), 10–16 (2021)

Chapter 13
Discussion on Preferable Probability, Discretization, Error Analysis, and Hybrid of Sequential Uniform Design with PMOO

Abstract Here in this chapter, the authors would like to write some words to initiate further discussion for the following problems. It includes discussion on the conception of preferable probability and its evaluation as well as discretized treatment, error analysis, and the hybrid of sequential uniform design with the probability-based multi-objective optimization for successive optimization, weighting factor. The authors wish this work will concrete a brick to attract jade and would make its contributions to relevant fields as a paving stone.

13.1 On Preferable Probability

In the practical process of material selection, it involves issues of "Simultaneous Optimization of Multi-Objective" of material performance in general. In the viewpoint of system theory, a material is a physical system containing both beneficial type of attributes and unbeneficial type of attributes inevitably in the selection, i.e., multiple objectives coexist inside the system. In respect of probability theory, the general operation for "simultaneous optimization of multiple indexes" is to take the form of "joint probability" rationally. So, the overall/integral event of "simultaneous optimization of multi-objective" is corresponding to the product of the probabilities of each individual objective (event) that appears at the same time, which thus transfers the multi-objective optimization problem into an overall (integrated) single-objective optimization problem logically. Therefore, in order to quantitatively describe the term "the higher the better" for the utility index of material performance indicator appropriately, one needs to seek aid from probability theory itself once more. The result is that one needs to transfer the common term of "the higher the better" for the utility index of material performance indicator into the language of probability theory quantitatively. Finally, a new concept of "preferable probability" is created to characterize the preference degree of utility index in the selection for a candidate,

The original version of this chapter was revised: Table 13.11 has been corrected. The correction to this chapter is available at
https://doi.org/10.1007/978-981-99-3939-8_15

M. Zheng et al., *Probability-Based Multi-objective Optimization for Material Selection*, https://doi.org/10.1007/978-981-99-3939-8_13

and furthermore, the direct and convenient hypothesis is that the partial preferable probability of the utility index with the character of "the higher the better" (beneficial type of index) is positively correlative to the value of the corresponding utility index linearly for the simplicity; equivalently the partial preferable probability of the utility index with the character of "the lower the better" (unbeneficial type of index) is negatively correlative to the value of the corresponding utility index in linear manner. The subsequent assessment of preferable probability can thus be conducted quantitatively. Through this procedure, one could establish an integrated methodology for the overall consideration of "Simultaneous Optimization of the Multiple Objectives" of material performance in the selection of the system in principle.

13.2 On the Assessments of Robustness of Performance Utility with Uncertainty

The continuing improvement and stability of product quality are attractive things in the world, which are extremely concerned by both manufacturers and customers.

In 1950s [1], Taguchi attempted to introduce the robust parameter design (RPD) methodology for improving quality of a product or process by optimizing it with less sensitive to uncontrollable factors. Meanwhile, the idea of signal-to-noise ratio (SNR) and its assessment were proposed by Taguchi [1]. Thereafter, many discussions were addressed to his methodology [2–5].

In fact, for sufficient number of test data the mean value of the test data $\overline{y}$ and the mean deviation s are the representative of the tested consequences in some sense. It was suggested by statisticians to take separate models to reflect the effects of both mean value $\overline{y}$ and variance s, i.e., $\overline{y}$ and s should be taken as independent responses for a set of actual experiments or processes in principle. Therefore, the optimization of both $\overline{y}$ closing to the target and minima of s should be treated with individual approaches in appropriate manners at the same time, as is done in Chap. 4 of this book.

However, Taguchi's adopted *SNR*, which concretes the two responses $\overline{y}$ and s prematurely into one index, induces insufficient optimization of the both $\overline{y}$ closing to the target and minima of s. Actually, the optimization of the maximum for SNR is not equivalent to the simultaneous optimization of both minima of s and $\overline{y}$ closing to its target individually.

Box further took two examples to show the difficulty of using SNR to decide the comparative preference of their consequences [2]. His two hypothetical samples are with four observed values in the circumstances of the smaller the better, respectively, which were,

The observed values of the **Hypothetical Example 1** are: 0, 0, 4, 4 $\rightarrow \overline{y} = 2$, $s = 2.31$; MSE = 8, SNR = − 9.03.

The observed values of the **Hypothetical Example 2** are: 1, 2, 3, 4 $\rightarrow \overline{y} = 2.5$, $s = 1.29$; MSE = 7.5, SNS = − 8.75.

Box argued that ***Example 1*** had smaller mean, but ***Example 2*** had smaller variance, smaller mean square error (MSE) and hence smaller quadratic loss around *0* and the more desirable value of SNR [2]. So, his problem was which example showed a preferable result. In his opining, the ***Example 1*** seems to exhibit clearly preferable to ***Example* 2** in the response value *0*, which represents perfection, while ***Example 1*** had two such kind of values, whereas ***Example 2*** had none such kind of value.

The probabilistic robustness assessment in Chap. 4 of this book takes both mean value $\overline{y}$ and variance s as two individual responses which follow the suggestion of statisticians to be as separate models, its rationality is logically appropriate.

As to this problem, in spirit of probability-based multi-objective optimization both the mean value $\overline{y}$ and variance s all belong to unbeneficial type of performances. Thus, the partial preferable probabilities of the mean value and the variance and the total preferable probability of Hypothetical Example 1 are 0.5556, 0.3583, and 0.1991, respectively; the partial preferable probabilities of the mean value and the variance and the total preferable probability of Hypothetical Example 2 are 0.4444, 0.6417, and 0.2852, individually.

Therefore, the ***Example 2*** is superior to ***Example 1*** in spirit of probability-based multi-objective optimization due to its larger total preferable probability.

Actually, the hypothetical (or observed) data in the above two examples are not too sufficient but scattered in the viewpoint of statistics. If the actual cases of experiments are really like this, supplement experiments should be done to provide more data so as to make the decision with sufficient test data.

13.3 On the Number of Discretized Sampling Points of Evaluation in Probability-Based Multi-objective Optimization by Means of GLP and Uniform Experimental Design

In Chap. 7 of this book, it illuminates roughly but not proves mathematically that 11, 17, and 19 sampling points are preliminarily appropriate for assessing complicated integral in one-, two-, and three-dimensional problems, respectively.

In fact, according to Hua and Wang [6] and Fang and Wang [7], Larcher and Traunfellner [8], and Fang et al. [9], as to the good lattice point (GLP), the discrepancy of the point set is $O(n^{-1}(\log n)^{s-1})$ for s—dimension with prime number n, so if we take 11 GLPs for one-dimensional case, the value of $O(1/11) \approx 0.0909$, i.e., less than 0.10%. Similarly, for two-dimensional condition, if we try to adopt 17 GLPs, its value of $O(n^{-1}(\log n)^{s-1})$ becomes $O(17^{-1}(\log 17)) \approx 0.0724$, which is close to the situation of one-dimensional case. While for three-dimensional condition, if we accept to employ 19 GLPs, its value of $O(n^{-1}(\log n)^{s-1})$ is approximately $O(19^{-1}(\log 19)^2)$ ≈ 0.0861, which is in the same magnitude order as that of the situation of one-dimensional case. Moreover if we adopt 23, 29, or even 31 GLPs for 3-D, the values for $O(n^{-1}(\log n)^{s-1})$ are 0.0806, 0.0737, and 0.0717, respectively, which are all in the same order of magnitude roughly.

In addition, according to Fang et al. [10], by using GLPs to generate an uniform design table $U_n(n^q)$, there are some constraints or rules for the creation. According to number theory [10], for each positive integer n, there exists a unique prime decomposition $n = p_1^{r1} \cdot p_2^{r2} \ldots p_t^{rt}$, where $p_1, p_2, \ldots, p_t$ present different primes and $r_1, r_2, \ldots, r_t$ are positive integers. Furthermore, the Euler function for this n is $\varphi(n) = n \cdot (1 - 1/p_1) \cdot (1 - 1/p_2) \ldots (1 - 1/p_r)$. Moreover, it proved that the number of possible columns q in uniform design table generated by GLPs is $q = \varphi(n)$, but the number of independent variables in the uniform design table to use is at most $\varphi(n)/2 + 1$, i.e., $s \leq \varphi(n)/2 + 1$ for each n [10], this is the constraint for using GLPs to create uniform design table. A design based on the lowest discrepancy in the space is close to a uniform design.

Figure 13.1 shows the variations of discrepancy for some independent variables s with respect to n, which is created from the data of Fang for uniform design table $U_n{}^*(n^q)$ [11]. It can be seen that the entire tendency of discrepancy for all s decreases with n, which is subjected to little wavy trace locally.

Besides, in the error analysis, we used the sine function $f(x) = B \cdot [1 + \mathrm{Sin}(2\pi x/C)]$ in Chap. 7, the assessment was conducted around the peak point, i.e., at $x = x_0 = C/4$; a general discussion for *one* dimension with 11 partitions is conducted here further in the following sections.

Fig. 13.1 Variations of discrepancy for some s with respect to n

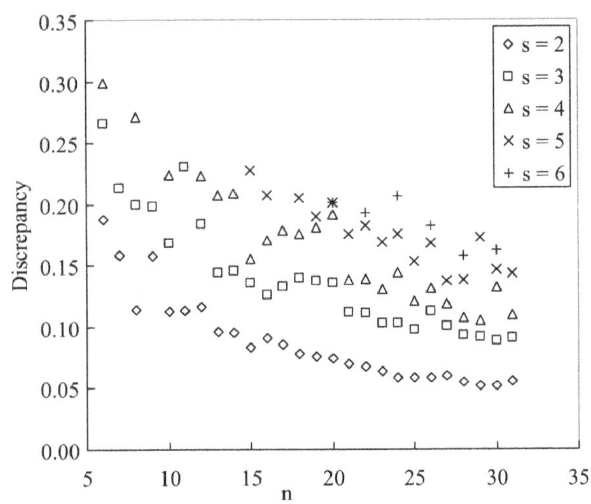

13.4 Error Analysis

(1) Error Analysis of Employing Uniform Design to Conduct Estimation of Definite Integral

In fact, under condition of one dimension, the distribution of GLPs is the same as that of the midpoint rule of the rectangular method for numerical integration [6, 7, 12–14].

In the midpoint rule, the integration $\overline{E(f)} = \int_a^b f(x) \cdot \mathrm{d}x$ is approximated by the summation of $\overline{f(D_n)} = \frac{(b-a)}{n} \sum_{i=1}^{n} f(a + (i - 1/2) \cdot (b - a)/n)$ with the total error of $E_M = \frac{1}{24} \frac{(b-a)^3}{n^2} \cdot \frac{1}{n} \sum_{i=1}^{n} f''(a + (i - 1/2) \cdot (b - a)/n)$ [12–14], f'' $(a + (i - 1/2) \cdot (b - a)/n)$ indicates the second derivative at location "$a + (i–1/2) \cdot (b - a)/n$". In the case of wavy integrand $f(x)$, the summation $S_n = \sum_{i=1}^{n} f''[a + (i - 1/2) \cdot (b - a)/n]$ itself is quite small; while for some monotonic function like e^x the summation S_n retains, even in this case, 11 sampling points could give good precious [15]. The integral $\int_0^3 \mathrm{e}^x \cdot \mathrm{d}x$ is with the accurate value of $\overline{E(f)} = \int_0^3 e^x \cdot dx = 19.0855$, while the summation of the discretized sampling points by means of midpoint rule is $\overline{f(D_{11})} = \frac{3}{11} \sum_{i=1}^{11} e^{[(i-1/2)\cdot 3/11]} = 19.0265$. This example together with those in Chap. 7 preliminarily indicates the applicability of the proposed approach in the viewpoint of practical engineering, of which the abstruse physical detail related to the spatial correlations of spatial sampling points are valuable to explore by mathematician [16].

As to higher dimensions, the total error could be estimated by $|E_M| \leq \frac{1}{24 \cdot s \cdot n} \sum_{l=1}^{s} (b_l - a_l)^3 \cdot M(l)$, $M(l)$ expresses the maximum of second partial derivative of $f(x)$ with respect to l-th variable $x_l \in [a_l, b_l]$, $\max f''(x)$.

Here, two further examples are provided to present that the approach of discretized sampling points with characteristic of GLP for assessing definite integrals is superior to Monte Carlo simulation.

(A) A comparison of Monte Carlo simulation and discretized sampling points with characteristic of GLP for assessing integral $f_1 \equiv \int_0^6 f_1(x) \cdot \mathrm{d}x = \int_0^6 (2 - x/3) \cdot \mathrm{d}x$.

Han et al. conducted the calculation of definite integral of $f_1 = \int_0^6 (2 - x/3) \cdot \mathrm{d}x$ by using Monte Carlo simulation [17]. The precious value of this integral is 6.

We could use 11 discretized sampling points with characteristic of GLP to restudy the assessment of this integration comparatively [15]. The positions of the GLPs for this problem are shown in Table 13.1 together with the discretized value of function $f_1(x)$.

According to the procedure in Chap. 7 of this book, as to the assessment by using 11 discretized sampling points, the definite integral is discretized as a summation. The summated value is $\frac{6}{11} \sum_{i=1}^{11} f_1(x_i) = 6$, which is exactly the same as that of the precious value of this integral. However, the error of the integral by using Monte Carlo simulation varies with the number of the sampling points even up to 1000 stochastic sampling points [17]. As to number of the sampling points of 100, the

Table 13.1 Positions of the GLPs together with the discretized value of function $f_1(x)$

No.	x	$f_1(x)$
1	0.2727	1.9091
2	0.8182	1.7273
3	1.3636	1.5455
4	1.9091	1.3636
5	2.4545	1.1818
6	3	1
7	3.5455	0.8182
8	4.0909	0.6365
9	4.6364	0.4545
10	5.1818	0.2727
11	5.7273	0.0909

error of the simulated value is 0.2908 with respect to the previous value of 6, and for the number of sampling points 1000 the error of the simulated value is 0.1214 [17], this consequence again indicates the applicability of the approach of discretized sampling points with characteristic of GLP for assessing the integral.

(B) A comparison of Monte Carlo simulation and discretized sampling points with characteristic of GLP for evaluating integral $f_2 \equiv \int_{x_1=1}^{2}\int_{x_2=1}^{x_1^2} f_2(x_1, x_2)\cdot \mathrm{d}x_1\mathrm{d}x_2 = \int_{x_1=1}^{2}\int_{x_2=1}^{x_1^2} x_1x_2\mathrm{d}x_1\mathrm{d}x_2$.

Li et al. performed the calculation of $f_2 \equiv \int_{x_1=1}^{2}\int_{x_2=1}^{x_1^2} f_2(x_1, x_2)\cdot \mathrm{d}x_1\mathrm{d}x_2 = \int_{x_1=1}^{2}\int_{x_2=1}^{x_1^2} x_1x_2\mathrm{d}x_1\mathrm{d}x_2$ by using Monte Carlo simulation [18]. The precious value of such integral is 4.5.

Here, lets' use the discretized sampling points with characteristic of GLP to reanalyze the assessment of this integration comparatively. The domain of the integral is in the area of triangle *ACD* of Fig. 13.2, and the corresponding virtual rectangle is the area *ABCD*. The ratio of the area of triangle *ACD* to the area of the corresponding virtual rectangle *ABCD* is 44.44%, so the $n_1 = 17$ sampling points sitting within area *ACD* uniformly is corresponding to the number of $n_0 = 17/0.4444 \approx 39$ sampling points filling in the virtual rectangle area *ABCD* evenly [6, 7].

Actually, from the comparative analysis, we could get a near prime number of $n_0\prime = 41$ for our usage to create an uniform design table [6, 7], and thus the distributions of the sampling points can be formed, which is shown in Table 13.2, while the actual number of the sampling points falling in the triangle area of *ACD* is fortunately 17. The discretized values of function $f_2(x_1, x_2)$ are supplied in Table 13.2 as well. The actual measure of the triangle area *ACD* is 4/3.

Again according to the procedure in Chap. 7, as to the assessment by using 17 discretized sampling points, the definite integral is discretized as a summation. The summated value for this problem is $\frac{4/3}{17}\sum_{j=1}^{17} f_2(x_{1j}, x_{2j}) = 4.5926$, which is

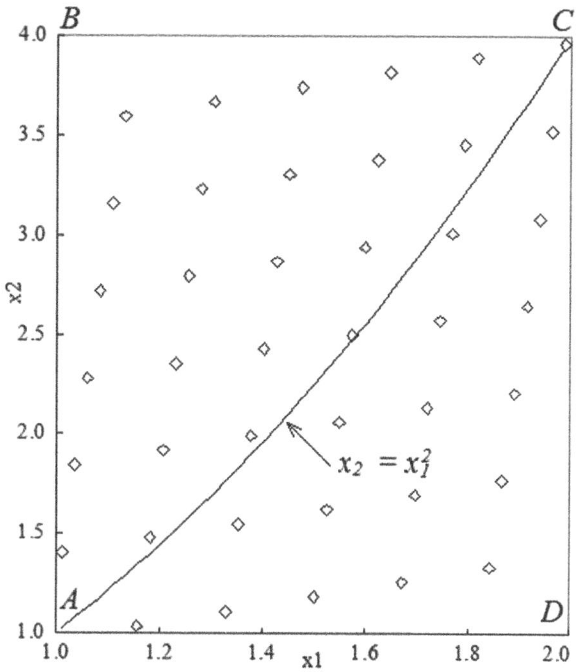

Fig. 13.2 Domain of the integral in area of triangle *ACD* and the corresponding virtual rectangle *ABCD*

with the error of 0.0926 with respect to the precious value of this integral of 4.5, correspondingly the relative error is about 2.06%.

However, the error of Li et al. by using Monte Carlo simulation exhibits significant oscillation till 1000 sampling points, and then the errors increase rapidly reaching about 0.19 at 1500 sampling points; thereafter, it decreases in oscillation manner [18]. This result shows that the approach of discretized sampling points with characteristic of GLP for assessing the integral is superior to Monte Carlo simulation obviously.

(2) Error Analysis of Employing Uniform Design to Estimate Maximum Value of a Function

In general, for a s—dimensional problem, if the function $f(\vec{x})$ is within a domain of $[0, 1]^s$, which takes maximum value $f(\vec{x}_p)$ at a discretized point $\vec{x}_p, \vec{x}_p \in [0, 1]$then the error between the actual maximum value of the function $f_{\max}(\vec{x})$ and the maximum value $f(\vec{x}_p)$ at a discretized point $\vec{x}_p$ is $E_{\text{actual}} = f_{\max}(\vec{x}) - f(\vec{x}_p)$, and estimation can be performed by following formula [15], $E_{\text{est}} \approx \frac{1}{2}\max\left|f(\vec{x}_p) - f(\vec{x}_{p+i})\right|$, or $E_{est} \approx \frac{1}{2\gamma}\sum_{i=1}^{\gamma}\left|f(\vec{x}_{p+i}) - f(\vec{x}_p)\right|$, $\vec{x}_{p+i}$ expresses the nearest neighbors of $\vec{x}_p$, γ is the number of the nearest neighbors of point $\vec{x}_p$.

Therefore, the error between the actual maximum value of the function $f_{\max}(\vec{x})$ and the maximum value $f(\vec{x}_p)$ at a discretized point $\vec{x}_p$ within its domain can be approximately estimated by using above formula in principle.

Table 13.2 Distributions of the sampling points with characteristic of GLP

No.	x_1	x_2	Notice	$f_2(x_1, x_2)$
1	1.0122	1.4024		
2	1.0366	1.8415		
3	1.061	2.2805		
4	1.0854	2.7195		
5	1.1098	3.1585		
6	1.1341	3.5976		
7	1.1585	1.0366	In ACD	1.2009
8	1.1829	1.4756		
9	1.2073	1.9146		
10	1.2317	2.3537		
11	1.2561	2.7927		
12	1.2802	3.2317		
13	1.3049	3.6707		
14	1.3293	1.1098	In ACD	1.4752
15	1.3537	1.5488	In ACD	2.0965
16	1.3780	1.9878		
17	1.4024	2.4268		
18	1.4268	2.8659		
19	1.4512	3.3049		
20	1.4756	3.7439		
21	1.5	1.1829	In ACD	1.7744
22	1.5244	1.6220	In ACD	2.4725
23	1.5488	2.0610	In ACD	3.1920
24	1.5732	2.5		
25	1.5976	2.9390		
26	1.6220	3.3780		
27	1.6463	3.8171		
28	1.6707	1.2561	In ACD	2.0986
29	1.6951	1.6951	In ACD	2.8734
30	1.7195	2.1341	In ACD	3.6697
31	1.7439	2.5732	In ACD	4.4874
32	1.7683	3.0122	In ACD	5.3264
33	1.7927	3.4512		
34	1.8171	3.8902		
35	1.8415	1.3293	In ACD	2.4478
36	1.8659	1.7683	In ACD	3.2994
37	1.8902	2.2073	In ACD	4.1724

(continued)

Table 13.2 (continued)

No.	x_1	x_2	Notice	$f_2(x_1, x_2)$
38	1.9146	2.6463	In ACD	5.0668
39	1.9390	3.0854	In ACD	5.9826
40	1.9634	3.5244	In ACD	6.9198
41	1.9878	3.9634		

13.5 Hybrid of Sequential Uniform Design with Probability-Based Multi-objective Optimization

In order to improve the accuracy of approximate maximum by using discretization method, sequential algorithm for optimization can be combined with the probability-based multi-objective optimization in its discretization.

(1) **Procedure**

Fang and Wang once proposed a sequential algorithm for optimization (SNTO) with NT-nets for uniform design in solving maximum value problem [7, 19–23], here we could develop a procedure or operation process for the hybrid of sequential uniform design with the probability-based multi-objective optimization to conduct further optimization analogously.

Assume SNTO be conducted within a rectangle domain of $D = [\boldsymbol{a}, \boldsymbol{b}]$. In our case, the maximum value of total preferable probability P_i can be assessed for the point set in each step. Thus, the operation process of SNTO algorithm for hybrid of sequential uniform design with the probability-based multi-objective optimization is as follows:

0th Step: Initialization.

At moment $t = 0$, $D^{(O)} = D$, $\boldsymbol{a}^{(O)} = \boldsymbol{a}$ and $\boldsymbol{b}^{(O)} = \boldsymbol{b}$.

1st Step: Generation of an NT—net.

The generation of a n_t points $\mathrm{P}(t)$ uniformly distributed on $D^{(t)} = [\boldsymbol{a}^{(t)}, \boldsymbol{b}^{(t)}]$ is conducted by using number-theoretic method. The sampling scheme in the point set has the maximum value of total preferable probability $P_i(\boldsymbol{x}^{(t)})$ at moment t.

2nd Step: Calculation of a novel approximate value.

Suppose $\boldsymbol{x}^{(t)} \in Ç^{(t)} \cup \{\boldsymbol{x}^{(t-1)}\}$ and $M^{(t)}$ such that $M^{(t)} = P_i(\boldsymbol{x}^{(t)}) \le P_i(\boldsymbol{y})$ for number of points with characteristic of $n_{t-1} = n_t = \ldots$, $\forall \boldsymbol{y} \in Ç^{(t)} \cup \{\boldsymbol{x}^{(t-1)}\}$, in which $\boldsymbol{x}^{(-1)}$ is the empty set, $\boldsymbol{x}^{(t)}$ and $M^{(t)}$ are the best approximations to $\boldsymbol{x}^*$ and M temporarily.

3rd Step: Termination condition.

Assume $c^{(t)} = (\text{Max } P_i^{(t-1)} - \text{Max } P_i^{(t)})/\text{Max } P_i^{(t-1)}$. If $c^{(t)} < \delta$, a pre-assigned small quantity, then $x^{(t)}$ and $M^{(t)}$ are acceptable; terminate algorithm. Otherwise, proceed to next step.

4th Step: Domain contraction.

A new domain can be set, i.e., $D(t + 1) = [\boldsymbol{a}^{(t+1)}, \boldsymbol{b}^{(t+1)}]$ can be formed as follows: $a_j^{(t+1)} = \max\,(x_j^{(t)} - bc_j^{(t)}, a_j)$ and $b_j^{(t+1)} = \min\,(x_j^{(t)} + bc_j^{(t)}, b_j)$, where b is a predefined contraction ratio. Set $t = t + 1$. Go to Step 1.

According to their experiences [7], Fang and Wang's suggestion is, $n_1 > n_2 = n_3 = \ldots$ for the processing. The contraction ratio b could be taken as 0.5. While Niederreiter and Peart [7] advised using $b_k = b^k$ as a contraction ratio at the k-th step with $b > 0$ as constant.

Remarks: in our case, at k-th step, $P_i(\boldsymbol{x}^{(k)}) \le P_i(\boldsymbol{x}^{(k-1)})$ in general for $k > 2$ only if $n_2 = n_3 = \ldots$. Or else, examine the domain contraction process again or stop the process of domain contraction, and take the $P_i(\boldsymbol{x}^{(k-1)})$ and the corresponding $\boldsymbol{x}^{(k-1)}$ as the optimal results.

(2) Application in Some Actual Problems

(A) **Robust Design of a Spring**

Take robust design of a spring with multi-objective optimization as an example.

The expression for the moment of force is $M = \frac{Ed^4}{3667Dn}\varphi$, in which n is the effective circles of the spring; ϕ is the rotational angle of the spring; E is the elasticity modulus of the spring, d is the wire diameter of the spring, and D is the middle diameter of the spring. The ranges of the variables are shown in Table 13.3.

The error ranges of variables are: $d \pm 0.046$ mm; $D \pm 0.2$ mm; $n \pm 2.78 \times 10^{-2}$; $\phi \pm 5°$. In addition, the desirable value of the moment of force is $M = 96$ N mm.

The uniform table $U_{37}(37^{12})$ is employed to conduct the discretization of this multi-objective optimization problem with four variables, and the design is shown in Table 13.4 from Fang's book [11]. The assessment results of the partial preferable probabilities for M and ΔM and the total preferable probabilities for each discrete point are presented in Table 13.5. The value of M is obtained by substituting parameters of Table 13.4 into the expression for M directly, and the value of ΔM is obtained by substituting parameters of Table 13.4 into the corresponding error transfer function of M. The error of M, i.e., ΔM, and deviation of M with respect to its desirable value 96 N mm ($\varepsilon \equiv |96 - M|$) all belong to the unbeneficial type of index, the assessments are shown in Table 13.5.

From Table 13.5, the maximum value of the total preferable probability is at the Test No. 11 with specific values of $x_1 = 0.9851$ mm, $x_2 = 6.0405$ mm, $x_3 = 3.1052$ and $x_4 = 33.0541°$, $M = 95.0493$ N mm and $\Delta M = 36.1288$ N mm, respectively. Furthermore, according to the procedure described in previous section, subsequent processing is used to contract the domain to conduct further evaluations.

The uniform table $U_{23}{}^*(23^7)$ from Fang's book is used to perform the successive assessments [12]. Table 13.6 shows the consequences of the successive evaluations.

Table 13.6 shows that the $c^{(k)}$ value at the 3rd step is 0.19%, if we set $\delta = 0.2\%$ as the pre-assigned small quantity for engineering application, then the final optimal consequences for this multi-objective optimization problem are $M = 95.2700$ N mm

Table 13.3 Ranges of the variables

Variable	d/mm	D/mm	n	ϕ /°	E/GPa
Range	[0.9, 1.2]	[6.0, 7.0]	[1.889, 3.889]	[24, 34]	210

Table 13.4 Design of the multi-objective optimization of spring problem with four variables due to $U_{37}(37^{12})$

No.	x_{10}	x_{20}	x_{30}	x_{40}	d/mm	D/mm	n	$\phi/°$
1	1	17	29	30	0.9041	6.4459	3.4295	31.9730
2	2	34	21	23	0.9122	6.9054	2.9971	30.0811
3	3	14	13	16	0.9203	6.3649	2.5647	28.1892
4	4	31	5	9	0.9284	6.8243	2.1322	26.2973
5	5	11	34	2	0.9365	6.2838	3.6998	24.4054
6	6	28	26	32	0.9446	6.7432	3.2674	32.5135
7	7	8	18	25	0.9527	6.2027	2.8349	30.6216
8	8	25	10	18	0.9608	6.6622	2.4025	28.7297
9	9	5	2	11	0.9689	6.1216	1.9701	26.8378
10	10	22	31	4	0.9770	6.5811	3.5376	24.9460
11	11	2	23	34	0.9851	6.0405	3.1052	33.0541
12	12	19	15	27	0.9932	6.5000	2.6728	31.1622
13	13	36	7	20	1.0012	6.9595	2.2404	29.2703
14	14	16	36	13	1.0095	6.4189	3.8079	27.3784
15	15	33	28	6	1.0176	6.8784	3.3755	25.4865
16	16	13	20	36	1.0257	6.3378	2.9431	33.5946
17	17	30	12	29	1.0338	6.7973	2.5106	31.7027
18	18	10	4	22	1.0419	6.2568	2.0782	29.8108
19	19	27	33	15	1.0500	6.7162	3.6458	27.9189
20	20	7	25	8	1.0581	6.1757	3.2134	26.0270
21	21	24	17	1	1.0662	6.6351	2.7809	24.1351
22	22	4	9	31	1.0743	6.0946	2.3485	32.2432
23	23	21	1	24	1.0824	6.5541	1.9160	30.3514
24	24	1	30	17	1.0905	6.0135	3.4836	28.4595
25	25	18	22	10	1.0986	6.4730	3.0512	26.5676
26	26	35	14	3	1.1066	6.9324	2.6187	24.6757
27	27	15	6	33	1.1149	6.3919	2.1863	32.7838
28	28	32	35	26	1.1230	6.8514	3.7539	30.8919
29	29	12	27	19	1.1311	6.3108	3.3214	29.0000
30	30	29	19	12	1.1392	6.7700	2.8890	27.1081
31	31	9	11	5	1.1473	6.2297	2.4566	25.2162
32	32	26	3	35	1.1554	6.6892	2.0241	33.3243
33	33	6	32	28	1.1635	6.1486	3.5917	31.4324
34	34	23	24	21	1.1716	6.6081	3.1593	29.5405
35	35	3	16	14	1.1797	6.0676	2.7268	27.6487
36	36	20	8	7	1.1878	6.5270	2.2944	25.7568
37	37	37	37	37	1.1959	6.9865	3.8620	33.8649

Table 13.5 Assessment results of the partial preferable probabilities for ε and ΔM and the total preferable probabilities of each discrete point

No.	M	$\varepsilon \equiv \lvert M - 96 \rvert$	$\pm \Delta M$	P_ε	$P_{\Delta M}$	$P_i \times 10^4$
1	55.3282	40.6718	22.0783	0.0268	0.0368	9.8578
2	57.6233	38.3767	23.4051	0.0273	0.0362	9.9004
3	70.9302	25.0699	29.7606	0.0304	0.0335	10.2012
4	76.8817	19.1183	33.1109	0.0318	0.0321	10.2137
5	46.2381	49.7619	20.3768	0.0247	0.0375	9.2595
6	67.2800	28.7200	26.0200	0.0296	0.0351	10.3851
7	82.1562	13.8438	32.7366	0.0330	0.0323	10.6591
8	87.6013	8.3987	35.6653	0.0342	0.0310	10.6422
9	112.3195	16.3195	47.5099	0.0324	0.0260	8.4455
10	55.9143	40.0857	23.8759	0.0269	0.0360	9.7030
11	95.0493	0.9507	36.1288	0.0360	0.0308	11.1081
12	99.9727	3.9727	38.6767	0.0353	0.0298	10.5111
13	108.0910	12.0910	42.7734	0.0334	0.0280	9.3708
14	66.6073	29.3927	26.8667	0.0294	0.0348	10.2249
15	67.3983	28.6017	27.9244	0.0296	0.0343	10.1563
16	114.1507	18.1507	42.1479	0.0320	0.0283	9.0614
17	121.5083	25.5083	45.7113	0.0303	0.0268	8.1211
18	154.7171	58.7172	60.2884	0.0226	0.0206	4.6621
19	79.36917	16.6308	31.0915	0.0324	0.0330	10.6757
20	94.14865	1.8514	38.3223	0.0358	0.0299	10.7113
21	96.80669	0.8067	40.6471	0.0360	0.0289	10.4292
22	171.8545	75.8545	63.7572	0.0186	0.0192	3.5686
23	190.0107	94.0107	72.1565	0.0144	0.0156	2.2491
24	110.0389	14.0389	42.4366	0.0330	0.0282	9.2911
25	112.2345	16.2345	44.4097	0.0325	0.0273	8.8765
26	116.7911	20.7911	47.6912	0.0314	0.0260	8.1513
27	207.5476	111.5476	75.0413	0.0103	0.0144	1.4879
28	109.3894	13.3894	39.6321	0.0331	0.0294	9.7267
29	129.6793	33.6793	48.6494	0.0284	0.0255	7.2599
30	133.6711	37.6711	51.4806	0.0275	0.0244	6.6939
31	163.4912	67.4912	65.7370	0.0206	0.0183	3.7680
32	251.1854	155.1854	88.6497	0.0002	0.0086	0.0190
33	149.3799	53.3799	53.4004	0.0238	0.0235	5.6126
34	152.6906	56.6906	55.7888	0.0231	0.0225	5.1985
35	185.3696	89.3696	70.4341	0.0155	0.0163	2.5301
36	196.0842	100.0842	76.8229	0.0130	0.0136	1.7724
37	147.0401	51.0401	49.6001	0.0244	0.0251	6.1325

Table 13.6 Consequences of the successive evaluations by using U_{23}*(23^7)

Step, k		0	1	2	3
Domain		[0.9, 1.2] × [6, 7.0] × [1.889, 3.889] × [24, 34]	[0.9, 1.08] × [6, 6.6] × [2.5, 3.5] × [27, 32]	[0.95, 1.04] × [6.15, 6.45] × [2.9, 3.4] × [28.5, 31]	[0.99, 1.04] × [6.22, 6.42] × [3, 3.26] × [28.7, 29.9]
Optimum location	x_1^*	0.9851	1.0213	1.0380	1.0302
	x_2^*	6.0405	6.4043	6.3652	6.3287
	x_3^*	3.1052	3.1739	3.1283	3.0735
	x_4^*	33.0541	30.3696	28.9891	28.7261
M		95.0493	93.0889	96.8032	95.2700
ΔM		36.1288	35.8195	37.7574	37.4705
Maximum of total preferable probability $P_i \times 10^3$		1.1108	2.9129	2.8676	2.8622
$c^{(k)} \times \%$				1.5552	0.1883

and $\Delta M = 37.4705$ N mm at $x_1 = 1.0302$ mm, $x_2 = 6.3287$ mm, $x_3 = 3.0735$ and $x_4 = 28.7261°$.

(B) **Design of Wall-Climbing Robot**

A wall-climbing robot was designed by Zhong et al. [24], the minimum mass (M), the minimum deformation (T_1), and the minimum maximum stress (T_2) were taken as the optimization objectives [24]. They employed response surface method and Pareto solution set to conduct the optimization and select the "final solution." There are three input variables x_1, x_2, and x_3, of which the ranges are, x_1: 432–528 mm; x_2: 54–66 mm; x_3: 9–11 mm, respectively. Specifically, they adopted MOGA multi-objective genetic optimization algorithm (NSGA-H) and the second-generation non-gene dominating genetic algorithm. Its initial population generation used 3000 samples, while each iteration needed 600 samples. The maximum allowable genetic algebra was 20 generations, the variation coefficient was set to 0.01, and the cross-coefficient was set to 0.98. The maximum allowed Pareto ratio is set to 60%, and the convergence stability is set to 2%. Finally, the first group was selected from the three candidate points, and the optimized mass was $M = 9.51$ kg, the minimum deformation was $T_1 = 0.53$ mm, and the minimum maximum stress was $T_2 = 6.55$ MPa as their "final solution."

In the following section, the probability-based multi-objective optimization together with the uniform design and sequential optimization are used to resolve this problem comparatively [25].

In Zhong's experimental design, they employed the combination of parameter sensitivity screening and Latin hypercube test [24], which produced 15 sample points, the results are cited in Table 13.7.

Subsequently, Table 13.8 shows the evaluation results of the preferable probability and total preferable probability of M, T_1 and T_2 values and the ranking at these 15

Table 13.7 Design and results excerpted from Zhong

No.	Width, x_1 (mm)	Length, x_2 (mm)	Thickness, x_3 (mm)	Mass, M (kg)	Max deformation, T_1 (mm)	Max stress, T_2 (MPa)
1	486.4	63.2	10.1	11.49	0.62	7.42
2	505.6	60.2	10.3	11.19	0.58	7.48
3	435.2	60.8	9.3	10.12	0.75	9.97
4	499.2	54.4	10.8	10.64	0.6	6.96
5	441.6	56.3	9.4	9.54	0.73	9.03
6	**524.8**	**55.2**	**9.2**	**9.71**	**0.52**	**6.54**
7	492.8	64.8	10.6	12.29	0.61	6.92
8	518.4	62.4	10.5	11.91	0.6	6.32
9	460.8	56.8	10.1	10.19	0.68	8.45
10	454.4	61.6	10.9	11.74	0.72	8.48
11	473.6	58.4	9.7	10.31	0.64	7.34
12	448.1	64	9.6	10.92	0.72	8.93
13	480.2	59.2	9.8	10.58	0.63	7.74
14	512.5	65.6	9.2	11.05	0.55	7.65
15	467.2	57.6	10.4	10.67	0.67	8.24

sample points. As can be seen from Table 13.8, the sixth sampling point has the largest total preferable probability P_i, and therefore the optimization point should be near the sampling point No. 6.

The regressed expressions for the total preferable probability P_i, M, T_1 and T_2 versus input variables x_1, x_2, and x_3 are as follows:

$$\begin{aligned} P_i \times 10^4 = & -18.7985 + 0.256928x_1 - 0.5451x_2 - 4.99662x_3 \\ & - 0.00022x_1^2 + 0.003651x_2^2 \\ & + 0.123263x_3^2 - 0.00058x_1x_2 \\ & + 0.000456x_1x_3 + 0.033007x_2x_3, \\ R^2 = & \, 0.9922 \end{aligned} \tag{13.1}$$

$$\begin{aligned} M = & \, 10.80453 + 0.056328x_1 + 0.050971x_2 - 5.09686x_3 \\ & - 6.3 \times 10^{-5}x_1^2 - 0.00088x_2^2 \\ & + 0.142866x_3^2 - 0.00022x_1x_2 \\ & + 0.002309x_1x_3 + 0.032387x_2x_3, \\ R^2 = & \, 0.9990 \end{aligned} \tag{13.2}$$

Table 13.8 Evaluation results of partial preferable probability and total preferable probability of M, T_1 and T_2 and ranking at the 15 sampling points

No.	Partial preferable probability			Total preferable probability	Rank
	P_M	P_{T1}	P_{T2}	$P_i \times 10^4$	
1	0.0626	0.0688	0.0699	3.0134	8
2	0.0644	0.0729	0.0694	3.2610	6
3	0.0709	0.0557	0.0498	1.9667	15
4	0.0678	0.0709	0.0735	3.5311	3
5	0.0744	0.0577	0.0572	2.4573	12
6	0.0734	0.0789	0.0768	4.4535	1
7	0.0578	0.0698	0.0738	2.9802	9
8	0.0601	0.0709	0.0786	3.3451	4
9	0.0705	0.0628	0.0618	2.7338	11
10	0.0611	0.0587	0.0616	2.2084	14
11	0.0698	0.0668	0.0705	3.2879	5
12	0.0661	0.0587	0.0580	2.2503	13
13	0.0681	0.0678	0.0674	3.1138	6
14	0.0653	0.0759	0.0681	3.3752	2
15	0.0676	0.0638	0.0634	2.7347	10

$$\begin{aligned} T_1 &= 4.837274 - 0.02588x_1 + 0.024731x_2 \\ &\quad + 0.330507x_3 + 2.26 \times 10^{-5}x_1^2 - 0.00021x_2^2 - \\ &\quad 0.01433x_3^2 + 2 \times 10^{-5}x_1x_2 \\ &\quad + 6.53 \times 10^{-5}x_1x_3 - 0.00078x_2x_3, \\ R^2 &= 0.9920 \end{aligned} \tag{13.3}$$

$$\begin{aligned} T_2 &= -16.7851 - 0.39669x_1 + 1.943064x_2 \\ &\quad + 13.84361x_3 + 0.000429x_1^2 - 0.00879x_2^2 \\ &\quad - 0.17798x_3^2 + 0.00049x_1x_2 \\ &\quad - 0.00775x_1x_3 - 0.11098x_2x_3, \\ R^2 &= 0.9503 \end{aligned} \tag{13.4}$$

Near the sampling point No. 6, sequential optimization can be conducted to implement deep optimization. Here, the uniform design table $U^*_{19}(19^7)$ from Fang's book is used for sequential optimization [12], and the results are shown in Table 13.9, the evaluation of P_i at each sampling point is conducted with the partial preferable probability of predicted values of M, T_1 and T_2 by using Eq. (13.2) through Eq. (13.4) at each discrete sampling point. The P_i of zero-th step is 4.4535. Table 13.9 shows that

Table 13.9 Evaluation results of sequential optimization with $U_{19}{*}(19^7)$

Step, k	Domain (mm^3)	Optimum "coordinate"			Max P_i $\times 10^4$	$c^{(k)} \times \%$
		x_1^*	x_2^*	x_3^*		
0	[432, 528]×[54, 66]×[9, 11]	524.8	55.2	9.2	4.4535	
1	[500, 528]×[54, 60]×[9, 10]	515.4737	54.1579	9.1316	1.7985	
2	[517, 528]×[54, 57]×[9, 9.5]	519.9474	54.0790	9.0658	1.6469	8.4292
3	[514, 528]×[54, 55.5]×[9, 9.25]	521.7368	54.0395	9.0329	1.5582	5.3859
4	[517, 528]×[54, 54.8]×[9, 9.12]	523.0789	54.0211	9.0158	1.5111	3.0227

the maximum total preferable probability P_i decreases slowly. If the pre-assigned small quantity δ is set as $\delta = 4\%$ for $c^{(k)}$, then this sequential optimization can be terminated at the 4th step. At this point, the optimized result is at $x_1^* = 523.0789$ mm, $x_2^* = 54.0211$ mm, and $x_3^* = 9.0158$ mm, and the corresponding optimum objective values are: mass $M* = 9.46$ kg, minimum deformation $T_1^* = 0.50$ mm, and minimum maximum stress $T_2^* = 6.14$ MPa. These values of M^*, T_1^* and T_2^* are evaluated by employing Eq. (13.2) through Eq. (13.4) at the optimized position of $x_1^* = 523.0789$ mm, $x_2^* = 54.0211$ mm and $x_3^* = 9.0158$ mm.

Obviously, the current optimum results of M^*, T_1^* and T_2^* are superior to the "final solution" given by Zhong's. Besides, the optimization process of the current approach is relatively simple.

13.6 On Weighting Factor

With the rapid increasing variety of materials available today, and their excellent characteristics, applications, advantages, and limitations, the selection of materials for engineering designs with a deep understanding of functional requirements for each individual component is needed in considering various significant attributes or criteria. The objectives or attributes or criteria include: mechanical properties, manufacturing properties (machinability, weldability, formability, castability, heat treatability, etc.), physical properties, magnetic properties, electrical properties, chemical properties, cost, product shape, environmental resistance, availability, fashion, market trends, recycling, cultural aspects, aesthetics, target group, etc. [26].

It has been known that the optimal selection of a material for an engineering design from a large number of candidate materials with some objectives (attributes, criteria) is a typical multi-objective optimization problem (MOO). It is needed to provide a rational methodology for decision-makers to perform material selection properly in considering a number of selection attributes and their interrelations though the issue of selection and decision are complex.

In the previous approaches of material selection, each decision in MOO methods possesses four indispensable parts, i.e., (a) alternatives, (b) objectives (criteria or attributes), (c) weighting factor or relative importance of each attribute, and (d) measures of performance of alternatives with respect to the attributes [26].

The weighting factor or relative importance of attributes takes part a very significant role in the decision-making process of these previous approaches [26–31], which determines the final result of the optimal selection completely in the "additive algorithm" in their approaches.

In fact, the weighting factor or relative importance of the attributes is more like a factor of the decision-maker's subjective preference in some sense, though it was labeled as "objective weighting factor" and "subjective weighting factor" for division [26–31].

It was further stated that the weighting factor to the criteria in the "subjective weighting factor" is assigned by the DM or designer, the importance of the criteria in objective methods is not by DM or designer [26–31]. The subjective method assigns the weighting factor of attributes on basis of preference information of attributes given by expert evaluation and their previous experience and the specific constraints of design uniquely, or preferences of designer [27].

The objective method gives the weighting factor of attributes on basis of known data of the problem "solely" [27]. The combination of both "subjective weighting factor" and "objective weighting factor" can be employed to conduct the optimal selection as well [27].

As to the "objective method," the determination of "objective weighting factor" is non-unique, there exist the mean weight (equal importance) method, entropy method, standard deviation method, criteria importance through inter-criteria correlation, preference selection index, integrated methods, etc. [26–31]. In fact, there is no sufficient provision of the "objective method" to illuminate the objectivity of the corresponding approach. In addition, the selection of scaling factor of normalization (denominator) for each attribute in the "additive method" and the corresponding weighting factor evaluation due to the different units of attributes is unreasonable or short of objectivity.

Although a lot of weighting techniques in MOO have been proposed previously, the determination of the importance of criteria in material selection has not been seen as well performed with rationality and objectivity.

In the standard deviation method, it considered the disperse property of attribute, and the possible interdependency of relative importance of attribute and its quantitative data [26–31]. Thus, the relative importance of attribute is seen to be proportional to the standard deviation of the corresponding attribute. However, since the units of attributes are different, a scaling factor of normalization (denominator) for each attribute is involved once more, which makes the standard deviation method problematic, too.

In the newly developed probability-based multi-objective optimization (PMOO), the preferable probability P_{ij} appears to be the unique representative indicator of the performance utility X_{ij} of j-th attribute of i-th alternative (candidate, scheme). So, the relative importance of attribute might be dependent on the scattering of preferable probability P_{ij}. Taking this in mind, it could develop a new evaluation of standard deviation method for weighting factor w_j on basis of PMOO in the following equation,

$$w_j = C_j/\left(\sum_{j=1}^{m} C_j\right), C_j = \left\{\sum_{i=1}^{n}\left(P_{ij} - \frac{1}{n}\right)^2/n\right\}^{0.5} \tag{13.5}$$

In Eq. (13.5), m is the number of objective (attribute), and n is the number of alternative (candidate, scheme).

As an application of Eq. (13.5), lets' study a material selection problem.

The example is a material selection problem with five alternative materials and four objectives (attributes), which is shown in Table 13.10. Now, the steps of the methodology proposed above are described as following procedures.

Step 1. *Fundamental data.*

The objective is to evaluate the five alternative materials of material selection for a product [28–30]. The objectives (attributes) include: tensile strength (*TS*), yield strength (*YS*), density (*D*), and corrosion resistance (*CR*). In this problem, *TS*, *YS*, and *CR* belong to beneficial type of attributes with preference of the higher the better. Density (*D*) belongs to the unbeneficial type of attribute with preference of the lower the better. Corrosion resistance (*CR*) indicates a relative value.

Step 2. *Evaluation of preferable probability.*

The evaluations of both partial preferable probability of each performance utility index and total preferable probability of each alternative without considering specific weighting factor are conducted, which are shown in Table 13.11, the ranking is shown as well.

Step 3. *Evaluation of weighting factor of attribute on basis of standard deviation of preferable probability.*

The evaluations of weighting factor of attribute on basis of standard deviation of preferable probability of attribute are conducted according to above-developed algorithm. The results for this problem are shown in Table 13.12. The evaluation of total preferable probability P_i with weighting factor w_j is evaluated by using $P_i = \prod_{j=1}^{m} P_{ij}^{w_j}$.

Table 13.10 Fundamental data of the example

Material	Material selection attributes			
	Tensile strength (MPa)	Young's modulus (GPa)	Density (g/cm^3)	Corrosion resistance
1	1650	58.5	2.3	0.5
2	1000	45.4	2.1	0.335
3	350	21.7	2.6	0.335
4	2150	64.3	2.4	0.5
5	700	23	1.71	0.59

Table 13.11 Partial preferable probabilities and total preferable probability of each performance utility index and each alternative without specific weight factor

	Partial preferable probability				Total preferable probability	
Material	P_{TS}	P_{YS}	P_D	P_{CR}	$P_i \times 10^3$	Rank
1	0.2821	0.2748	0.1925	0.2212	3.3012	2
2	0.1709	0.2132	0.2117	0.1482	1.1438	3
3	0.0598	0.1019	0.1638	0.1482	0.1481	5
4	0.3675	0.3020	0.1830	0.2212	4.4928	1
5	0.1197	0.1080	0.2490	0.2611	0.8404	4

Table 13.12 Evaluations of weighting factor of attribute on basis of standard deviation of preferable probability

Parameter	TS	YS	D	CR
C_j	0.1111	0.0828	0.0290	0.0447
w_j	0.4153	0.3093	0.1083	0.1671

Table 13.13 Evaluation result of ranking with weighting factor

Material	P_i	Ranking
1	0.2578	2
2	0.1829	3
3	0.0916	5
4	0.2946	1
5	0.1430	4

Step 4. *Evaluation of ranking with weighting factor.*

The evaluations of ranking with weighting factor of each performance utility index are conducted, which are shown in Table 13.13.

It can be seen from the comparison of Tables 13.11 and 13.13, though the final rankings of both consequences are the same by chance, the details are with intrinsic differences. However, irrational approaches obtain different results [29–31].

13.7 Conclusion

The discussions in this chapter focus on the conception of preferable probability and its discretization, error analysis, the hybrid of sequential uniform design with the probability-based multi-objective optimization for successive optimization, and weighting factor, which aims to initiate further study in relevant areas so as to get more efficient results.

References

1. K. Dehnad, *Quality Control, Robust Design, and the Taguchi Method* (Wadsworth & Brooks/Cole Advanced Books & Software Pacific Grove, California, 1989)
2. G. Box, Response. Technometrics **30**(1), 38–40 (1988)
3. G. Box, S. Jones, Split-plot designs for robust product experimentation. J. Appl. Statist. **19**, 3–26 (1992)
4. G.E.P. Box, W.G. Hunter, J.S. Hunter, *Statistics for Experimenters* (Wiley, New York, 1978)
5. V.N. Nair (Ed.), Taguchi's parameter design: a panel discussion. Technometrics **34**, 127–161 (1992)
6. L.-K. Hua, Y. Wang, *Applications of Number Theory to Numerical Analysis* (Springer-Verlag & Science Press, Berlin; New York, Beijing, 1981). ISBN: 9783540103820.
7. K.-T. Fang, Y. Wang, *Number—Theoretic Methods in Statistics* (Chapman & Hall, London, 1994). ISBN: 0-412-46520-5
8. G. Larcher, C. Traunfellner, On the numerical integration of Walsh series by number—theoretic methods. Math. Computat. **63**(207), 277–291 (1994). https://www.jstor.org/stable/2153574
9. K.-T. Fang, D.K.J. Lin, P. Winker, Y. Zhang, Uniform design: theory and application. Technometrics **42**(3), 237–248 (2000). https://www.jstor.org/stable/1271079
10. K.T. Fang, M.Q. Liu, H. Qin, Y.D. Zhou, *Theory and Application of Uniform Experimental Design* (Science Press & Springer Nature, Beijing & Singapore, 2018)
11. K.T. Fang, *Uniform Design and Uniform Design Table* (Science Press, Beijing, 1994)
12. M.H. Holmes, *Analysis Introduction to Scientific Computing and Data Analysis* (Springer Nature, Cham, 2016)
13. J. Izaac, J. Wang, *Computational Quantum Mechanics* (Springer Nature, Cham, 2018)
14. B.A. Stickler, E. Schachinger, *Basic Concept in Computational Physics*, 2nd edn. (Springer Nature, Cham, 2016)
15. M. Zheng, J. Yu, H. Teng, Y. Wang, Error analysis for designed test and numerical integral by using UED in material research. Tehnicki Glas. **17**(2), 167–171 (2023). https://doi.org/10.31803/tg-20220504103756
16. B.D. Ripley, *Spatial Statistics* (John Wiley & Sons, NJ, 1981). ISBN: 0-47169116-X
17. J. Han, W. Ren, Monte Carlo integration and quasi-Monte Carlo integration. J. Shanxi Normal Univ. (Nat. Sci. Ed.) **l21**(1), 13–17 (2007)
18. M. Li, H. Wand, G. Zhang, Calculation of double integrals based on Monte Carlo method. J. Hainan Normal Univ. (Nat. Sci. Ed.) **23**(3), 242–244 (2010)
19. Y. Zhu, On the convergence of sequential number-theoretic method for optimization. Acta Math. Appl. Sin. **17**(4), 532–538 (2001)
20. F. Gong, H. Cui, L. Zhang, Y. Liang, An improved algorithm of sequential number—theoretic optimization SNTO/based on clustering technique. Chemom. Intell. Lab. Syst. **45**, 339–346 (1999)
21. Y.B. Ji, G. Alaerts, C.J. Xu, Y.Z. Hu, Y.V. Heyden, Sequential uniform designs for fingerprints development of *Ginkgo Biloba* extracts by capillary electrophoresis. J. Chromatogr. A **1128**, 273–281 (2006)
22. Q. Xu, Y. Liang, Z. Hou, A multi-sequential number—theoretic optimization algorithm using clustering methods. J. Cent. South Univ. Technol. **12**(S1), 283–293 (2005). 1005-9784(2005)S1-0283-11
23. M. Zheng, H. Teng, Y. Wang, Hybrids of uniform test and sequential uniform designs with "intersection" method for multi objective optimization. Tehnicki Glas. **17**(1), 94–97 (2023). https://doi.org/10.31803/tg-20211130132744
24. Z. Zhong, F. Liu, T. Sun, C. Di, Multi-objective optimization design of wall-climbing robot based on response surface method. Mach. Electron. **40**(6), 36–40 (2022). 1001-2257(2022)06-0036-05
25. M. Zheng, J. Yu, A probabilistic multi-objective optimization mechanical design. J. Umm Al-Qura Univ. Eng. Archit. **14**, 115–121 (2023). https://doi.org/10.1007/s43995-023-00023-3

26. R.V. Rao, B.K. Patel, A subjective and objective integrated multiple attribute decision making method for material selection. Mater. Design **31**, 4738–4747 (2010)
27. A. Jahan, F. Mustapha, S.M. Sapuan, M.Y. Ismail, M. Bahraminasab, A framework for weighting of criteria in ranking stage of material selection process. Int. J. Adv. Manuf. Technol. **58**, 411–420 (2012)
28. D. Diakoulaki, G. Mavrotas, L. Papayannakis, Determing objective weightings in multiple criteria problems: the critic method. Comput. Ops. Res. **22**, 763–770 (1995)
29. K. Maniya, M.G. Bhatt, A selection of material using a novel type decision-making method: preference selection index method. Mater. Design **31**, 1785–1789 (2010)
30. A. Jahan, F. Mustapha, M.Y. Ismail, S.M. Sapuan, M. Bahraminasab, A comprehensive VIKOR method for material selection. Mater. Design **32**, 1215–1221 (2011)
31. R.V. Rao, A decision making methodology for material selection using an improved compromise ranking method. Mater. Design **29**, 1949–1954 (2008)

Chapter 14
General Conclusions

Abstract This chapter gives a summary of the whole book, mainly the intrinsic idea of probability-based multi-objective optimization and its application.

Through thorough description in this book, the authors would like to comprehensively summarize the following terms:

1. The probability-based multi-objective optimization reflects the essence of simultaneous optimization of multiple objectives appropriately in the viewpoint of system theory in overall respects.
2. Each utility of material performance indicator can be characterized by the newly developed conception of partial preferable probability quantitatively; the total preferable probability is the unique representative of each candidate material.
3. The algorithm for assessing partial preferable probability *vs* utility of material performance indicator is based on the function and preference of the material performance indicator in the optimization.
4. The probability-based treatment for robustness (or interval number) is consistent with the essence of simultaneous optimization for the arithmetic mean value and the variance of a utility data with dispersed value appropriately.
5. The treatment for the multi-objective optimization containing desirable response is formulated properly.
6. The hybrids of probability-based multi-objective optimization with experimental designs are conducted, which are brand new methodologies.
7. The discretization in assessing probability-based multi-objective optimization is based on GLP and uniform design, in which the distribution of the sampling points is well deterministic and uniform according to the rules of GLP and uniform design.
8. At to the material itself, the service condition, the processing technology and the whole cost in its entire lifetime are needed to be considered by means of utilities of material performance indicators.
9. Treatments of portfolio investment and multi-objective shortest path problem by means of probability-based multi-objective optimization are with rationality.

M. Zheng et al., *Probability-Based Multi-objective Optimization for Material Selection*, https://doi.org/10.1007/978-981-99-3939-8_14

10. The fuzzy-based probabilistic multi-objective optimization and cluster analysis of multiple objectives are preliminarily formulated.
11. Further exploration of developments and applications in more fields is expected and with bright prospects. This work is aimed to cast a brick to attract jade in the future and would make its contributions to relevant fields as a paving stone.

Correction to: Probability-Based Multi-objective Optimization for Material Selection

Correction to:
M. Zheng et al., *Probability-Based Multi-objective Optimization for Material Selection*,
https://doi.org/10.1007/978-981-99-3939-8

The original version of the book was inadvertently published with an incorrect Equations. These chapters 6, 11 and 13 corrections have been amended. The correction chapters and book have been updated with the changes.

The updated original version of these chapters can be found at
https://doi.org/10.1007/978-981-99-3939-8_6
https://doi.org/10.1007/978-981-99-3939-8_11
https://doi.org/10.1007/978-981-99-3939-8_13

M. Zheng et al., *Probability-Based Multi-objective Optimization for Material Selection*, https://doi.org/10.1007/978-981-99-3939-8_15

The manufacturer's authorised representative in the EU is Springer Nature Customer Service Centre GmbH, Europaplatz 3, 69115 Heidelberg, Germany. If you have any concerns regarding our products, please contact ProductSafety@springernature.com

Printed and bound by CPI Group (UK) Ltd, Croydon, CR0 4YY

30/09/2026

02225314-0004